The item should be returned or renewed by the last date stamped below.

Dylid dychwelyd neu adnewyddu'r eitem erbyn y dyddiad olaf sydd wedi'i stampio isod

PILLGWENLLY

To renew visit / Adnewyddwch ar
www.newport.gov.uk/libraries

Fruit

Essential know-how and expert advice for gardening success

CONTENTS

Elderberries and greengages are often extremely hard to find in the shops, but with a little thought and planning, these delicious fruits are easy to grow at home in your garden.

PLANNING, PLANTING, AND GROWING

Whether you have a large garden or a window box, it's easy to grow fruit. This section helps you choose the right fruit for you and your plot and gives the practical advice you'll need to enjoy abundant harvests from healthy plants.

PLANNING YOUR FRUIT GARDEN

Take some time to plan what you'd like to grow and the best place to grow it; determine which fruits will be best suited to you and your plot – be it a garden, allotment, in containers, or even in your house. New dwarf varieties of fruit that previously took a lot of space to grow are being introduced all the time, so now even the smallest area can be productive. Most fruit trees and bushes are relatively low-maintenance, but consider how much time you can give to looking after your plants before you buy them. It's always a good idea, especially if you're new to gardening, to start with just a few plants or a small patch and build up your plot gradually, until you've got the hang of what to do and how long it will take you.

PLANTING FOR SUCCESS

Investing some time into making informed choices about fruit varieties, preparing the soil, and planting correctly will pay dividends for many years to come. Avoid mistakes by learning about rootstocks, pollination groups, soil acidity, harvest periods, and heritage varieties. You'll soon be buying fruit like a seasoned expert by learning what to look for and when to buy it; by knowing if you should be getting a potted or a bare-root plant; and by understanding the fundamentals of good pruning.

YOUR CORNER OF THE PLANET

Gardening is a terrific way to connect with nature and enjoy spending quality time outdoors, so don't let your fruit trees and bushes become a battlefield. Learn to work with nature and to share your space with wildlife, and you'll find a happy balance, whereby everyone and everything gets to enjoy your garden and its bounty. Fruit blossom is a boon for pollinators such as bees, and many species will find a home among your trees and bushes. Feed and water your plants in environmentally friendly ways, while reducing your fruits' food miles to zero. Many plants produce so much fruit that it's impossible to eat it all fresh, but by preserving it you'll be able to eat homegrown fruit all year round.

YOUR FRUIT GARDEN

There are numerous reasons for wanting to grow your own fruit. You may want more immediate and direct control over your food origins and how your food is grown; perhaps you're keen to save some money, or to have access to fruits that you can't buy easily in the shops. Whatever the reasons you want to grow – and no matter how big or small your space – everyone can grow some fruit in their garden. Start off by growing a few of your favourite fruits, and then expand from there if you want to, so that your fruit garden is always a pleasure and never a chore.

A fruit tree in the garden introduces the next generation to the delights of picking fresh fruit.

Prolific redcurrants need only a few hours of maintenance a year.

Grow fruit such as strawberries and rhubarb in raised beds if reaching high or low is a problem for you.

EASY TO GROW

Growing fruit, unlike vegetables, has an undeserved reputation for being complicated and high-maintenance. Nothing could be further from the truth. A fruit tree or bush, once planted, will provide you with harvests (sometimes of many kilos) year after year in return for minimal attention. Pruning fruit trees and bushes is straightforward and even if you do make a mistake, it's not the end of the world – they'll just keep growing. Fruit plants can be trees, bushes, climbers, or small perennials, and with the many types of trees and bushes available, there are fruit plants suitable for every size and style of garden.

TOP TIP TO GROW FRUIT IN A MORE SOCIABLE WAY, SEEK OUT ALLOTMENTS, COMMUNITY GARDENS, OR GARDEN-SHARE SCHEMES AS PLACES YOU CAN GARDEN AND MEET PEOPLE.

FRESH FRUIT AND FRESH AIR

Homegrown fruit is delicious for so many reasons: it's a variety with characteristics that you prefer; it's fresh, ripe, and in season; and, as something you've grown yourself, it's wonderfully satisfying to eat. Don't get caught up in traditional ideas of self-sufficiency: instead, focus on growing the fruit that suits you, your garden, and your lifestyle and circumstances. Having a fruit garden needn't be hard work, but it does offer good, gentle exercise. And there are other physical benefits to getting your hands dirty, as fresh air and contact with the soil give a huge boost to our immune system.

An allotment can be a place of escape but it also offers the chance to join a community of like-minded people.

Damsons are rarely seen in shops, but they're a nutritious and delicious fruit.

INCREASING WELLBEING

Your garden might be a place of glorious solitude, but it can also be a place to meet with your family, friends, and neighbours. Either way, gardening has been proven to elevate mood and alleviate stress and fatigue. A garden is also a place to interact with nature and a place to be creative – both have great benefits for mental health. Growing, especially growing food, allows you to connect with like-minded people, and to swap tips and harvests: a sense of community is also terrific for our happiness and wellbeing. And buying plants from local nurseries is a sustainable choice for the environment as well as an opportunity to have a good chat with knowledgeable growers.

WHY GROW FRUIT?

Whether you have a big garden or a few pots on a balcony, it's relatively easy to grow your own fruit – and there's nothing quite like picking a handful of fresh raspberries for breakfast from right outside your door, making jam from plums you've grown yourself, or being able to store enough homegrown apples to last all winter. Once you grow your own fruit, you'll wonder how you ever managed without it.

Fresh, homegrown fruit harvests bring a greater connection with the seasons.

EATING FRESH

One of the greatest benefits of growing your own fruit is that you can eat it fresh from the plant at its ripest and most succulent and delicious. Both the flavours and the sweetness of a fruit develop until it's fully ripe; but they start to degrade as soon as the fruit is picked. The fruit you buy in shops is often picked underripe to preserve it through the days, weeks, or months that it will be stored, transported, and then displayed – all of which makes it less tasty and of less nutritional value.

Raising your own fruit means you can choose to grow organically (without artificial or chemical fertilizers and pesticides), so you know exactly what you're eating. It also allows you to eat seasonally, with all the health benefits derived from fruit that's in its prime.

Bare-root fruit bushes and trees are less expensive than potted specimens.

GOOD INVESTMENT

Growing your own fruit requires a small initial outlay in money and time to buy your trees or bushes and then plant them. However, after that, all they'll need is minimal annual maintenance. For example, for the price of a few punnets of blackcurrants and a little bit of your time spent in the garden, you could buy a bush that will amply reward you with kilos of fruit every year for many years to come.

Celebrate the harvest with toffee apples made from scrumptious, fresh, homegrown fruit.

HELPING THE PLANET

Planting just one fruit tree or bush in your garden can make a positive difference to the natural world. The plants themselves absorb carbon dioxide from the atmosphere, thereby helping to offset carbon emissions. By eating homegrown rather than shop-bought you'll reduce your food miles, as well as the packaging and petrochemicals (present in commercial fertilizers and pesticides) used in food production, at immense cost to the environment.

TOP TIP BIRDS LOVE FRUIT AND BERRIES. ENCOURAGE THEM INTO YOUR GARDEN AND YOUR AREA BY LEAVING SOME FRUIT ON THE PLANTS FOR THEM TO FEED ON.

Growing more plants also means you'll attract birds, bees, and other wildlife – many of which are threatened species – and help to support them by providing food and shelter (see pp.42–43). Gardening also enables us to gain a closer connection with nature and the seasons, boosting our wellbeing.

There are many more cherry varieties available to grow than there are to buy as fruit in the shops.

GREATER CHOICE

Growing your own fruit gives you a huge choice – not only of types of fruit (such as greengages, mulberries, quinces, and persimmons) but also of the many varieties of those fruits, each with their own attributes and flavour profiles.

Commercial growers have different priorities from home growers, and tend to use only a few varieties. Farmers regularly supply supermarkets, and must take account of how well their fruit will survive through being picked, stored, and transported. Strawberry seeds, for example, are naturally dimpled into the fruit skin, making it bruise easily. Plant breeders have developed varieties with the seeds borne on virtually smooth and thus firmer flesh so that the berries are less likely to bruise and rot in the punnet. Unfortunately, focus on this, and on making the fruit uniformly large, has resulted in loss of flavour. But growing your own means you can enjoy traditional tastes such as 'Cambridge Favourite', with its sunken seeds.

Bees and other pollinators benefit from the enormous number of flowers a blossoming fruit garden provides.

WHICH FRUIT TO GROW?

When starting a fruit garden, it's easy to get carried away with all the delicious possibilities. However, it's best to start small and then expand your list of fruits once you have a clear idea of what's involved and how much time and effort will be required. Visiting local public kitchen gardens or allotment open days can be a great source of inspiration for deciding what will work best in your garden.

Figs are ideal for growing in containers in a sunny courtyard.

Strawberries are tastiest when picked perfectly ripe and warmed by the sun.

GROW FOR YOUR KITCHEN

Concentrate on growing fruits that you and your family love to eat and cook with, so that you can make your own desserts, jams, and cordials from homegrown harvests (see p.46). If possible, select different varieties to spread the harvests over the season, or to explore the fabulous different flavour profiles within one type of fruit.

Some fruits, such as peaches and plums, especially benefit from being picked when ripe from the tree or bush for a superior flavour to shop-bought, although this is true of most fruit. It's also good to grow fruits you love that aren't easily available in shops, such as fresh gooseberries and blackcurrants.

GROW FOR YOUR GARDEN AND LIFESTYE

Fruit will grow in almost any situation, but choosing what will thrive, rather than merely survive, will guarantee success. Your choice of what to grow will be guided by practicalities, including what will suit your garden and lifestyle.

Consider the size and situation of your plot and what plants would suit it best (see pp.14–15). Do you want to grow trees, bushes, climbers, or all three? Do you have vertical space that could be used for trained trees, grape vines, or climbers? Do you need to

Even the walls of your house can be used to grow fruit, trained onto wires or trellis, as with this pear tree.

grow plants in pots or do you have a lawn or border you can turn over to a fruit garden – or perhaps a mix of both?

Think also about budget. Fruit plants require more initial financial outlay than a vegetable patch, because you have to buy plants rather than seeds, but once you've bought them, they'll reward you at harvest time for years to come.

NEED TO KNOW

All fruit is quite straightforward to grow, but the easiest and lowest-maintenance of all are: Autumn-fruiting raspberry (pp.128–129) • Blackcurrant (pp.100–101) • Elder (pp.80–81) • Quince (pp.54–55) • Rhubarb (pp.98–99) • Strawberry (pp.94–95)

New dwarf varieties available from plant breeders mean that most fruit can now be grown in containers, even large plants like blackberries.

WHERE TO GROW

If their roots are in the soil and their heads in the sun, most fruit plants will be happy. The range of dwarf and trained-form options for even the biggest fruit plants gives you tremendous flexibility over where and how to grow your fruit garden; but paying attention to fundamentals such as sun, soil types, and exposure means you'll be giving your plants the best possible situation.

Most allotment associations allow fruit bushes, but check whether you can also plant trees.

Fruit trees need sunshine to ripen and develop the sweetness of their fruit.

Hanging baskets of strawberries make the most of both sunny spots on walls and your garden space.

SUN OR SHADE?

The more sunshine fruit trees and bushes receive, the more likely it is that they'll bear fruit. Bees and other insects work best in the warmth of the sun, ensuring your fruit blossom is pollinated properly; plants need the sun to create their own energy through photosynthesis. Therefore, if possible, site your plants in a sunny spot. Growing trained fruit such as plums and pears on a south- or southwest-facing wall is ideal. If your back garden is shady, think about whether you can train fruit against the front wall of your house. Lifting the height of plants can also help them get more sun – strawberries, for example, can grow in hanging baskets; or consider growing your fruit in raised beds.

SOIL TYPES

The primary concern with the type of soil in your garden is how well it drains. Most fruit trees and bushes prefer a soil that's not too wet and not too dry, with plenty of nutrient-rich organic matter (see pp.26–27). Make sure plants aren't waterlogged in winter (they'll essentially drown if sitting in water). All soils can be improved by adding lots of organic matter through mulching (see p.26). If your soil is poor, consider adding a raised bed over the top of it to improve the richness and drainage of the top layer.

Some fruits, such as blueberries and cranberries, need acidic soil to grow well. Acidity (and its opposite, alkalinity) is measured using the pH scale. Simple home kits for testing your soil pH are widely available at garden centres and online: it's worth testing your soil before growing acid-loving plants (follow the

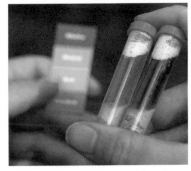

It's easy to measure the acidity of your garden soil with a home test.

instructions on the kit). If your soil is too alkaline, you can add acidifying products such as sulphur chips to the soil, but it's an expensive option that must be repeated annually. A better option is to grow these plants in raised beds or containers filled with specialist acidic ("ericaceous") compost (see pp.18–19).

Blueberries require acidic soil (a low pH score) to grow well.

FROST AND WIND

Frost is the enemy of many fruit species because it can freeze their spring blossom. The freezing and defrosting damages the flowers, preventing them from developing into fruit and resulting in a poor harvest. Avoid planting in frost pockets (areas where frost reliably settles on cold nights). If your garden is in the colder north, choosing hardier

varieties will help; you may wish to grow some fruits under cover of a greenhouse or polytunnel (see pp.32–33). Strong and cold winds can harm blossom and fruit (causing them to fall), and desiccate the soil and plants – in such cases, you'll need to water more often. If your garden is exposed, consider planting a windbreak evergreen hedge (more effective and better for wildlife than a wall or fence) to protect your plants.

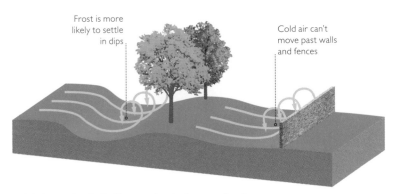

Frost is more likely to settle in dips

Cold air can't move past walls and fences

Cold air settles in hollows and at the bottom of walls and fences, so frost is more likely here.

An evergreen hedge slows down strong winds, protecting your garden.

FRUIT IN SMALL SPACES

You don't need a vast garden to grow fruit. The walled kitchen gardens of stately homes are inspirational in this respect, and well worth a visit for the ingenuity with which space is utilized – training fruit trees on the walls, for example, or along the edges of beds. So almost every space, however small, can be put to use, be it a courtyard, balcony, window box, or even just your front doorstep.

A lot of pear varieties will fit into a small space when grown as cordons along a sunny wall.

TRAINED TREES

Training trees as cordons (single or double vertical or diagonal stems), espaliers (many horizontal stems), or fans (several stems fanning out from the trunk diagonally), essentially reduces their depth to a few centimetres. They can therefore be planted against walls or fences and take up very little space. Alternatively, put in some posts with wires between them and train the trees onto the wires to create a living screen (see p.86; and p.18 for fruit in containers).

This apple espalier at the back of a border adds extra, edible screening above a low fence.

> **TOP TIP** IF YOU HAVE ONLY ONE SPACE FOR A TREE BUT WOULD LIKE MORE THAN ONE KIND OF FRUIT, OR NEED TWO VARIETIES FOR POLLINATION, THEN PLANT TWO OR MORE TREES IN THE SAME HOLE (IT'S BEST TO PLANT TREES OF A SIMILAR SIZE AND SHAPE). PRUNE SO THAT THE BRANCHES ARE DIRECTED OUTWARDS. YIELDS FROM INDIVIDUAL TREES WILL BE LOWER, BUT THE PRODUCTIVITY OF THE SPACE WILL BE MAXIMIZED.

FOREST GARDENING

With edible "forest gardening", you can fit a lot of plants into a small space by using shade-tolerant fruits and planting them closer together. This replicates the canopy layers (trees, bushes, and ground cover) of a woodland or forest on a small scale with edible plants. For example, try planting a fruit tree (such as a pear, apple, or cherry) with currant bushes beneath it, and alpine strawberries to cover the ground.

Blackcurrants will grow well in the dappled shade under fruit trees in an edible forest garden.

Many fruits, including strawberries, thrive in pots; some can even be trained on walls.

FRUIT IN CONTAINERS

Even if you don't have soil to grow your plants in, it's possible to have a bountiful fruit garden grown entirely in containers. Courtyards, balconies, or pots around your front door can all be used to grow fruit bushes and trees successfully. And for those who are lucky enough to have a garden, it can still be useful to grow some fruit in pots. The size and drainage capabilities of containers are more important than the material they're made from, so use whatever works for you and your budget – be it upcycled, second-hand, or brand new.

Protect early-blossoming peaches by growing them in pots and moving them to a sheltered spot in spring.

A fig tree will fruit well in a container on a sunny patio.

WHAT TO GROW?

Most fruits can be grown in a container, provided it's big enough. Fruit trees specifically marketed as "patio" or "ballerina" trees are a good choice, or get your chosen variety on a dwarfing rootstock (see p.24). There are also varieties of cane fruits and bushes that have been bred for containers, such as the raspberry 'Ruby Beauty'. Strawberries and cranberries can be grown in smaller pots and even hanging baskets.

ADVANTAGES OF POTS

Aside from the aesthetic value of a potted fruit garden, there are a number of situations when it may be preferable to grow fruit in a container – not least of which is that you can easily take them with you if you move house.

Some fruits, such as blueberries and cranberries, need acidic soil that may be easier to supply in a container than the ground. Restricting the roots of fig trees in a pot promotes fruiting over foliage growth. You might not be able to plant into the ground at the base of walls or a pergola, but by planting into a pot you can still train plants such as grapes up these structures.

Growing in containers also allows you to move plants such as citrus, peaches, and apricots to protect them or their early blossom from cold weather; tender passion fruit and pineapple can be grown under cover all year in pots.

Consider using pots and containers whenever it's not otherwise possible to grow – for example, by creating a strawberry patch using just window boxes and hanging baskets.

Blueberry bushes are ideal to grow in pots, but make sure you add drainage holes to upcycled containers.

Harvest the last fruits before the first frost or snowfall or they'll have to be left for the birds.

Winter pruning on a sunny afternoon is not just more pleasant for you, it's healthier for the tree.

WINTER

Once the last of the fruits have been harvested and any rotten, fallen fruits and diseased leaves cleared away, the only remaining task is the winter pruning of fruit trees and bushes. You can do this at any time between the last of the leaves falling in autumn and the first of the buds breaking in spring. This is when the plants are dormant, conserving their energy in preparation for spring – which isn't a bad idea for the gardener either.

FRUIT HARVEST CALENDAR

With some careful choices of fruits and varieties, you can be picking fresh fruit from your garden from early spring to late autumn. This table shows the whole harvest season for each of the fruits in the directories; individual varieties will be ready to harvest in the early, middle, or late period of that season.

	SPRING			SUMMER			AUTUMN			WINTER		
Apple						■	■	■	■			
Apricot					■	■						
Blackberry					■	■	■					
Blackcurrant					■	■						
Blueberry					■	■						
Cherry				■	■							
Chilean guava							■	■	■			
Citrus fruits	■	■	■	■	■	■	■	■	■	■	■	■
Cranberry							■	■	■			
Elder		■	■			■	■					
Fig						■	■					
Goji berry						■	■	■				
Gooseberry				■	■							
Grape							■	■				
Hybrid berries					■	■	■					
Kiwi							■	■				
Lingonberry						■	■	■	■			
Medlar									■	■	■	
Melon						■	■	■				
Mulberry					■	■						
Passion fruit						■	■	■				
Peach, nectarine					■	■						
Pear							■	■				
Persimmon							■	■	■	■	■	
Pineapple	■	■	■	■	■	■	■	■	■	■	■	■
Plum, gage, damson						■	■	■				
Quince								■	■	■		
Raspberry					■	■	■	■	■			
Redcurrant, whitecurrant					■	■	■					
Rhubarb	■	■	■	■	■							
Strawberry				■	■	■	■					

CHOOSING A VARIETY

Once you've decided which fruits you want to grow, the next step is to choose the variety that meets your needs. Some varieties will be better suited than others to the light conditions, microclimate, and soil in your garden. Some will crop over a long period, while others may produce a glut.

A few varietes have specific pollination requirements (*see p.24*). Apart from these practicalities, choice of variety is a matter of taste: for example, do you like crunchy, nutty apples, or juicy, sweet ones? The directories (*see pp.50–135*) list some of the tastiest and most reliable fruit varieties.

Planting autumn fruiting varieties, such as raspberry 'All Gold', which bears delicious large yellow fruit, will extend your growing season.

EXTENDING THE HARVEST

From spring rhubarb to autumn quince, a range of different plants will supply you with fresh fruit for months. Making clever use of the many varieties of fruit will also help you extend your growing year and harvest the right fruit at the right time. For example, try growing varieties of strawberries classed as "everbearers" (which provide regular crops from midsummer to autumn) with "summer bearers", which provide larger one-off harvests – perfect for making jams and preserves. Protecting early fruiting varieties under cloches as they come into flower can even give you a harvest in spring.

GROW FOR PURPOSE

Some fruit varieties are better suited to eating fresh and others to cooking (this is especially true for apples and plums), while some store or juice better than others. All varieties of all fruits taste different. Choose varieties depending on how you're going to use them – do you want to make your own preserves, eat fruit straight from the plant, or both? Garden shows can be a great opportunity to try different fruit varieties before you invest in plants.

Grow culinary plums for tarts and ice-cream, and dessert plums for eating fresh.

HERITAGE AND
LOCAL VARIETIES

The demand for year-round fruit and
the dominance of supermarkets have
led to the near-ubiquity of imported
fruits bred for longevity rather than
flavour. Meanwhile, many traditional
heirloom varieties have been ignored,
with a resultant loss of genetic diversity.
Choosing to grow heritage or heirloom
plants helps to preserve historic
varieties, which are often far tastier
than mass-produced fruits. Heritage
fruits are often specific to regions, so
try selecting ones that are (or were)
grown in your part of the country.
Specialist fruit growers are the best
sources for these.

Heirloom 'Forelle' pears originated
in Germany in the 1600s.

TOP TIP REFER TO ORGANIZATIONS SUCH AS SLOW FOOD AND THE
NATIONAL FRUIT COLLECTION FOR INFORMATION ON ENDANGERED AND
LOCAL HERITAGE VARIETIES, INCLUDING DAMSON 'SHROPSHIRE PRUNE'.

Community apple days can be a chance to discover new varieties,
and get your apples pressed into juice.

PLANT NAMES
EXPLAINED

"Variety" is a broad term often used
to describe the different "types" of
one species, but you may encounter
a number of terms that have specific
meanings to fruit growers.

VARIETY These are naturally
occurring variants of one species
identified with separate names, such
as the apple 'Bramley's Seedling'.

CULTIVAR This is short for "cultivated
variety", and denotes a variety that
has been created through selective
breeding. Today there is little
difference between a cultivar and
a variety and the two terms are
often used interchangeably.

HYBRID When two plants of different
species are successfully cross-bred,
they create a new plant that has
characteristics from both parents.
Such fruit hybrids are often crosses
between two similar fruits that
create a new type – for example,
the many berries bred from
raspberries and blackberries.

TRADE NAME Strict naming rules for
plant breeders can result in new
cultivars having two names – the
official name and the trade name.
The official name is denoted by the
abbreviation "PBR" (Plant Breeders'
Rights). The trade name has more
market appeal than the official name:
for example, Lingonberry Fireballs,
rather than 'Lirome' PBR.

AWARD OF GARDEN MERIT (AGM)
The UK's Royal Horticultural Society
tests a range of varieties. Those that
perform reliably and well are granted
an Award of Garden Merit (AGM).

CHOOSING THE RIGHT TREE

Rootstocks and pollination groups may sound complicated, but in fact they're extremely easy to navigate. To select a tree that will be adequately pollinated and the correct ultimate size for your garden, simply refer to the information supplied in the chart below and in the relevant plant entry (see pp.52–81). For less commonly used rootstocks, seek advice from a reputable nursery.

This 'Oullins Gage' flower (see p.75) needs a tree in the same pollination group nearby to be fertilized successfully and form a greengage.

ROOTSTOCKS

A rootstock is the root system of a tree that's used as a base upon which to grow another tree. The two are joined by means of grafting, a process that hasn't changed significantly since ancient times (see pp.48–49). While it is possible to grow fruit trees from seed, the result is usually disappointingly inferior, unmanageably vigorous, and not the same variety as the parent tree. Some rootstocks are used to confer a predictable level of vigour, or lack of it, and sometimes also disease resistance into the whole plant (such as the MM rootstocks for apples).

Use dwarfing or semi-dwarfing rootstocks for trained forms – the smaller you want the tree, the more dwarfing the rootstock should be. So with apples, for example, an M27 rootstock is better for a cordon; for an espalier or small tree you could use M26; and a freestanding tree on an M25 rootstock could reach 25m (82ft) high.

COMMON ROOTSTOCKS

Fruit	Rootstock	Vigour
Apple	M27	very dwarfing
Apple	M9	dwarfing
Apple	M26	semi-dwarfing
Apple	MM106	semi-vigorous
Apple	M111	semi-vigorous
Apple	M25	vigorous
Pear, quince	Quince C	semi-dwarfing
Pear, medlar, quince	Quince A	semi-vigorous
Plum	Pixy	semi-dwarfing
Plum, peach, apricot	St Julien A	semi-vigorous
Cherry	Tabel	semi-dwarfing
Cherry	Gisela 5	semi-dwarfing
Cherry	Colt	semi-dwarfing

On this apple tree, the rootstock and variety union point can be seen where the stem changes colour and direction.

POLLINATION GROUPS

Different fruit trees (even of the same fruit) flower at different times of the year, and need a nearby tree of the same type to pollinate the blossom. Although self-pollinating varieties don't need a compatible pollinating tree in the vicinity, they'll fruit better if there is one available.

If your tree requires a pollination partner, you'll need to plant two trees with pollination groups that are ideally the same but can be adjacent. For example, to grow apple 'Blenheim Orange', which is in pollination group 3, you'll also need an apple in pollination group 2, 3, or 4.

Beware though, as some varieties of pears, plums, and cherries are incompatible with other varieties in the same group – for example, both 'Doyenne du Comice' and 'Onward' pears are in pollination group 4 but will not self- or cross-pollinate themselves or each other. None of the varieties listed in the directories in this book are incompatible with each other.

Crab apple trees can act as pollinators for apples flowering at the same time, and dandelions will attract bees to your garden.

PREPARING AND MAINTAINING THE SOIL

Most fruit plants are perennials, growing for many years – once planted, they'll be in the ground for the foreseeable future. Before planting, take the opportunity to improve the quality of your soil: it's easy to do. If you feed it every year, your plants will grow healthy and strong, without the need for artificial fertilizers; use homemade compost, both to save money and to help the environment.

Look after your soil and it will repay you year after year.

Use a spade to mix compost into the soil if you don't have a garden fork.

PREPARE TO PLANT

Before planting, remove any weeds by digging out their roots; if lifting an area of lawn, slice horizontally with a spade under the grass in sections. (Pile up the sections of turf, upside down, ensuring no grass is peeping out: it will rot into lovely topsoil). Spread a thick layer of compost or organic matter over the area and use a garden fork to turn it into the ground. This will aerate the soil and relieve compaction, improving drainage. Tread down the bumps afterwards and rake level before digging your planting hole.

MULCHING

Over time, soil nutrients are washed away by rain and used up by plants, so replace them by regular mulching. Mulch is anything applied around the base of plants over the top of the soil. Using a mulch of compost or organic matter conserves soil moisture levels, suppresses weeds, and adds nutrients. Over the year, it will be incorporated into the soil by worms. Remove any weeds before they set seed; get the roots out too as they can resprout.

Add mulch as a thick layer (about 8cm/3in) around the base of the plants every late winter. The mulch shouldn't be touching stems or trunks as it might induce rot. Add extra nutrients by mixing in wood ash (high in potassium, see p.37) or a proprietary biochar product (carbonized organic material).

Mulch potted plants annually if they're not being repotted.

Making leaf mould is just speeding up nature's own recycling process.

COMPOST AND LEAF MOULD

If you can, make your own compost as it's cheaper and avoids the plastic packaging associated with buying in bags. To keep your compost heap healthy, turn it regularly to aerate it (watch out for toads and slowworms that can hibernate in the heap) and add a balanced mix of greens (such as weeds, vegetable peelings, and grass clippings) and browns (such as dry leaves, cardboard, and small twigs).

Make the most of autumn leaves: water them if they're dry and then seal them into old compost or bin bags, piercing a few air holes in the side. Put them aside for two years and they'll turn into leaf mould, a wonderful compost or mulch.

Good compost is known as "black gold" because it's so valuable for the soil.

TOP TIP LARGER STEMS AND BRANCHES THAT HAVE BEEN PRUNED FROM FRUIT TREES CAN BE CHOPPED VERY SMALL AND ADDED TO YOUR COMPOST HEAP, OR WOVEN BETWEEN POSTS TO MAKE A WILDLIFE-FRIENDLY "DEAD HEDGE". ALTERNATIVELY, YOU CAN TAKE THEM TO YOUR MUNICIPAL RECYCLING CENTRE FOR COMPOSTING.

NEED TO KNOW
- Compost is rotted plant matter, either made in your own heap or bin, or bought in bags.
- Soil-based (or loam-based) composts are best for long-term planting in pots; topsoil can also be purchased to fill raised beds.
- Ensure your compost is peat-free – it's much better for the environment.
- Organic matter is any well-rotted material that can be applied to the soil to improve it, such as horse manure, leaf mould, or compost. It's important that organic matter is well-rotted (brown and crumbly), otherwise it will remove nutrients from the soil while it's still decomposing.

BUYING FRUIT PLANTS

There are numerous varieties of fruit available, so it's a good idea to prepare a list before shopping – this means there's less chance of being tempted into other purchases. Specialist fruit nurseries are the best places to buy quality fruit trees and bushes, and they'll help you with expert advice, both in person and online. Bear in mind that plants grown on site at local nurseries will be used to a similar climate and soil to that of your garden. Alternatively, you can pick up bare-root trees and potted bushes inexpensively in supermarkets and garden centres.

These potted peach trees are ready to be planted.

WHAT TO LOOK FOR

When buying plants, look for healthy specimens. If they're in leaf, the foliage should be green and abundant, with no signs of pests (check under the leaves too) or diseases (see pp.38–41). Trunks, stems, and branches should have no damage to the bark, no wounds or tears. The roots on bare-root plants should be unbroken and without wounds. Potted plants should have a well-developed root system, with some but not lots of loose compost in the pot; avoid rootbound plants with roots wound around the inside of the pot and with little compost left. Don't be afraid to take the plant out of the pot in the shop to check it. If buying online, make sure the retailer has a no-quibbles returns policy.

Bare-root trees need rehydrating in a bucket of water before planting.

POTTED OR BARE-ROOT?

Between autumn and late winter, many species are available to buy "bare-root" (with their roots bare), which is a relatively inexpensive way to buy trees. "Rootballed" plants will also have been dug up that season and can be bought through the winter but have then had their roots tightly wrapped in hessian or plastic (see right). At other times of year, plants growing in pots are available. This is the most expensive way to buy them, but for some species, it's the only way. The smaller the plant, the longer it will take to bear a decent crop, but the cheaper it will be.

Rootballed plants, like these, are less expensive than potted plants, but their roots can get damaged when they're being wrapped.

> **TOP TIP** TRANSPORT YOUR TREES INSIDE THE CAR, IF POSSIBLE (OR GET THEM DELIVERED). TO AVOID DAMAGE, IF THEY'RE GOING ON A ROOF-RACK OR IN A TRAILER, WRAP THEM SECURELY IN SHEETING.

Visit specialist fruit nurseries for good advice and the best range of varieties.

PLANTING

Planting is a task you'll only ever do once, so it's worth taking the time to make sure you get it right. Autumn is always the best season to plant: the soil is warm, and there's usually enough rain to mean that you won't have to water your new fruit trees and bushes very often. Otherwise, take care to avoid waterlogged and frozen soil during the winter months, and be prepared to water frequently when planting in spring and summer.

Blackcurrants are the only fruit that's planted below the height of the soil level.

Soak bare-root plants before planting to rehydrate them.

BEFORE PLANTING

If necessary, prepare the ground by weeding it thoroughly and turning in some compost or organic matter (see pp.26–27). It's better, but not crucial, to do this a couple of weeks in advance to give the soil time to settle. Put in training wires and posts if essential for your chosen plant.

Water potted plants thoroughly, and unwrap and soak bare-root plants in a bucket of water for an hour or so before planting them. Detach the plants from any canes and remove all remaining ties.

PLANTING FRUIT

Dig the hole just before planting rather than in advance to prevent the soil from drying out (or the hole from filling with water, depending on the weather). Use a garden fork to loosen very compacted soil at the base and sides if necessary. Whether a bush or tree is bare-root or potted, the method is the same: ensure the root flare (where the stems turn into roots) is at the same height as the soil level. You may have to scrape back some compost on potted plants to find the flare point. The hole needs to be deep and wide enough to accommodate all the roots, but no bigger than that.

Put in upright stakes at this point (see *opposite*). Holding the plant at the right level if needed, backfill around the roots with the soil, firming it in well; use your heel once all the soil is in (see *pp.138–139* for planting climbers and next to walls or fences).

Keep offering the plant up and making the hole bigger until it's the right width and depth.

Staking trees protects them from breaking or being uprooted in strong winds.

STAKING

Freestanding trees (in the ground and in pots) need a stake initially to protect them and ensure their trunks don't grow at a wonky angle. Pyramid trees need a permanent 2.5m (8ft) upright stake. Bare-root trees can be given a 1.2m (4ft) upright stake before planting; but for potted trees, put in a 1.5m (5ft) stake angled at 45° to the ground after planting to avoid damaging the rootball. Ensure there's around 60cm (24in) of stake driven into the ground to make it secure. Smaller trees and plants can be supported by bamboo canes or hazel poles, but make sure there's plenty of the stake in the ground to support the plant in high winds.

NEED TO KNOW

- Various tree ties are available, but style is not important if it works.
- A tie forms a buffer between the stake and the plant; some have a sponge for extra cushioning.
- A good tie won't slip or be too tight or too loose and should be easily adjustable as the tree grows.
- Garden twine is biodegradable but needs replacing at least annually.
- If using twine, fasten the plant to the stake or wire in a figure of eight shape, looping around both the stem and the stake/wire, and tying off against the stake/wire.
- Check and adjust ties regularly through the year.

Water newly planted trees even if rain is forecast.

Use a figure of eight tie when securing thinner stems to canes with twine.

AFTER PLANTING

After planting (and staking) your tree or bush, water it thoroughly; give it at least a full can's worth of water, but apply it gradually so that it sinks in rather than running off the soil surface.

Then spread a 5cm (2in) layer of mulch (compost, or well-rotted organic matter) around the base of the plant, ensuring it doesn't touch the stem(s). Add a label, or make a note on a plan of the garden of the variety name, along with its location.

GROWING UNDER COVER

It's possible to have a bountiful fruit garden outside with no need for a greenhouse, polytunnel, or conservatory – but some fruits appreciate being protected from the worst of the winter cold and wet. Others are tropical plants that simply won't survive outside in a temperate climate. If you can only grow inside your home because you have no outside space, some fruit will thrive as houseplants.

Alpine and standard strawberries can be grown in pots and troughs on a sunny windowsill.

Some fruit, like peaches, will crop better in pots under cover than outdoors.

PLANTS TO GROW UNDER COVER

Some fruits need to be, or benefit from being, grown under cover for all or part of the year (see pp.50–135).

TENDER FRUIT (Tropical) Need constant protection in temperate climates: Passion fruit • Pineapple

FRUIT THAT WILL CROP WELL Thrive under cover, but still crop outdoors: Apricot • Fig • Grape • Melon • Peach and nectarine

FRUIT TO BRING IN FOR WINTER Need protection from the cold and wet: Apricot • Citrus (essential) • Fig • Peach and nectarine

WHERE TO GROW

For protecting trees from winter cold, use an unheated greenhouse, conservatory, or polytunnel; it helps, but isn't crucial, if it's well-lit. If the only appropriate space you have is in the house, keep trees away from drying heat sources such as radiators.

If you're growing plants that crop well inside (see left), consider keeping them, if possible, in a conservatory or polytunnel all year round, especially in cooler, northern areas. The conditions are warmer here, so plants come into leaf sooner in spring, grow faster, and generally fruit better than those outside, although you may need to hand-pollinate some of your crops. Greenhouses need not be expensive: you may find an inexpensive or free one in your local listings if you're able to dismantle it and move it to your house.

Fruits that are suitable as houseplants if placed by large, sunny windows or in conservatories include strawberries and all plants listed left (see individual directory entries, pp.50–135, for further information; see pp.18–19 for advice on varieties suitable for pots).

A small greenhouse insulates tender and half-hardy fruit plants in cold weather.

NEED TO KNOW

- It's important that plants grown indoors receive sufficient water and nutrients (see pp.36–37).
- A number of pests can affect indoor and greenhouse plants, such as scale insects, mealybugs, and aphids (see pp.38–39).

Grapes provide both shade and a healthy crop on this conservatory rooftop.

PRUNING BASICS

Pruning might sound complicated, but it's really a very straightforward process of simply cutting some of the growth from a plant. It's a task that's necessary for various reasons: to keep the plant to a certain size or encourage a certain shape; to help maintain its health by increasing the air flow through it; and to promote flowering and fruiting growth rather than foliage. Refer to the pruning sections of the directories for more information on how to prune different fruit trees and bushes.

When summer pruning trained fruit such as this gooseberry cordon, cut to just above a leaf.

TOOLS

If you're growing fruit trees or fruit bushes, you'll need some good cutting tools for your annual pruning. Get the best quality you can afford, look after them well, and they should last a lifetime. Some manufacturers offer refurbishing services. A pair of secateurs will do most of the pruning of fruit bushes and young trees; the bypass style – whereby the blades cross, like a pair of scissors (see *right*) – is preferable to anvil secateurs (where one blade presses into the flat edge of the other blade). Loppers should be used to cut any branch thicker than your thumb – and again, bypass is better than anvil. A pruning saw is necessary for larger branches; but depending on what you plan on growing, you might not need one for a year or two, if ever. Long-handled loppers and saws are useful for taller trees, or a sturdy ladder.

Keep your tools clean and sharpen them regularly to ensure your cuts are hygienic and tidy. After pruning out diseased branches – and before using them on a healthy plant – it's important to clean your tools with hot soapy water to avoid spreading infection.

Protect your hands while cleaning and sharpening tools by wearing gloves.

TOP TIP THE CARPENTRY ADAGE OF "MEASURE TWICE, CUT ONCE" CAN BE USEFULLY ADAPTED TO FRUIT PRUNING: "LOOK TWICE, CUT ONCE". ALWAYS ASSESS THE OVERALL SHAPE, STRUCTURE, AND HEALTH OF YOUR PLANT AND DECIDE ON THE NECESSARY CUTS BEFORE EMBARKING ON ANY PRUNING.

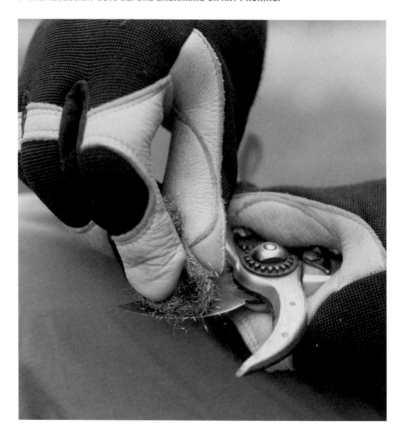

CUTTING SMALL BRANCHES

When cutting small branches, always prune to just above a bud (or a shoot, or leaf if it's summer). Cut too far away from the bud and the little stub between the cut and the bud will die off, leaving it vulnerable to infection by disease as well as looking ugly. Cut too close to the bud and the bud itself will be damaged – it won't be able to grow and that section of branch will die off. Make a clean cut using the right-sized tool for the job (either secateurs or loppers). If the cut leaves a ragged tear, sharpen your tools and make a new cut above the next bud down the branch to remove the damaged area.

Protect your skin from the weeping sap when pruning fig trees in summer.

This branch has been correctly pruned to just above a bud.

CUTTING LARGER BRANCHES

Larger branches that need sawing off your fruit tree should be cut in three separate stages (see below) to avoid the weight of the branch causing it to fall before it's cut all the way through, and ripping off the bark in the process. First, cut a short way through the branch from the underside to prevent this from happening. Then cut a little way further out from the trunk so that you can take the weight off the branch before making the final cut next to the trunk neatly. Make this final cut at a slight angle, avoiding cutting into the "collar", the circle of growth around where the branch joins the trunk, so the tree can easily heal the wound.

First, cut from underneath, about 30cm (12in) from the trunk, to about a quarter of the way through the branch.

Second, a little further away from the trunk than the first cut, saw all the way through the branch.

Finally, saw the branch stub off the trunk at an angle just out from the "collar" (where the branch adjoins the trunk).

WATERING AND FEEDING

Plants need only sunlight, water, and sufficient nutrients to thrive: supply these and your fruit garden will be bountiful and healthy. The more time you spend in the garden, the better you'll get to know your plants and garden conditions. Being mindful of these aspects will help you become more responsive to your plants and water and feed them effectively and sustainably.

Water established fruit trees growing in the ground only in very dry spells or droughts, as rain supplies the rest of the water they need.

Give all plants a good drink immediately after planting to help the soil settle around the roots and to ensure they have plenty of water.

WATERING WISELY

Watering is one of the gardener's most essential tasks. To avoid wasting water, always test the soil or compost by feeling it, to a finger's depth, before watering. Looks can be deceiving – a dry crust on the surface could conceal wet compost beneath, and light rain can dampen the surface but not penetrate to the roots.

It's important to water your newly planted and young trees and bushes until they're able to establish their roots into the ground; in very dry spells; and when they're flowering and fruiting. Containers need regular watering; and even during winter, don't let the compost dry out completely. Conversely, make sure the compost doesn't become waterlogged in very wet weather. Raising containers on bricks or pot "feet" (available from garden centres) helps drainage.

TOP TIP WILTING LEAVES MAY NOT BE A SIGN OF A WATER-STRESSED PLANT, PROVIDED THE SOIL OR COMPOST IS MOIST. PLANTS REDUCE WATER LOSS ON VERY HOT DAYS BY DELIBERATELY WILTING (A PROCESS KNOWN AS "SUN WILT") AND WILL PERK UP AGAIN IN THE COOLER EVENING.

Closed-top water butts are preferable to open-top ones, which are a hazard to wildlife (and children), lose water to evaporation, and attract midges and mosquitos.

SAVING WATER

Installing a water butt or two in your garden will save resources and be better for your plants. Using grey water (after washing up, for example) is another good way to reduce water use, but make sure you're using an eco-friendly washing up liquid and apply it to plants only every now and then to prevent soap building up in the soil or compost.

A watering can gives you more control than a hose and makes you mindful of the volume of water you're applying. Water slowly and thoroughly and allow each application to sink in before adding more; watering in the cooler morning or evening reduces evaporation loss.

FEEDING PLANTS

The best way to feed plants in the ground is to add a thick mulch of compost or other organic matter to the soil every year (see p.26), which will get incorporated over time by the worms. Add a layer of mulch to the top of pots and raised beds, but plants in these will also need liquid fertilizer as their roots can't extend as far to take in nutrients. But don't overfeed them, as this can cause toxins to build up in the compost; lead to excessive foliage growth at the expense of fruit production; and tempt in pests to the sappy new growth.

Mulching with compost after planting and every year feeds the soil around roots, conserves moisture, and suppresses weeds.

TYPES OF FERTILIZER

If you mulch your soil regularly, plants in the ground won't need any additional granular or pelleted fertilizer. Pot-grown plants require regular feeding during the growing season (see *directory entries, pp.50–135, for how often to feed*). A liquid feed as you water them is best for this, diluted according to label instructions. Proprietary organic liquid feeds include options such as seaweed extract. Alternatively, make your own by steeping a bucketful of comfrey or nettle leaves in water for up to a week. Strain and use the liquid as a fertilizer (diluting in a ratio of 1:10) and add the leaves to the compost.

When adding fertilizer to water, follow dosage instructions on the label.

NEED TO KNOW

- The three main nutrients a plant needs are nitrogen (N), phosphorus (P), and potassium (K). In an inorganic fertilizer, the ratios of these nutrients are shown on the label as N:P:K.
- Nitrogen promotes healthy, green foliage growth.
- Phosphorus helps the plant develop a healthy root system.
- Potassium is essential to the plant's immune system and promotes flowering and fruiting.
- Plants also need trace elements such as magnesium and iron.

FRUIT PESTS

Fruit trees and bushes are generally less troubled by pests than vegetables are, but there are a few issues to look out for. An obvious problem is wasps and birds eating fruit, but this is a small loss that's far outweighed by the great work they do predating on many other pests. Always consider physical solutions rather than resorting to chemical sprays.

APHIDS (WOOLLY)

PROBLEM Fluffy, white, waxy growth on bark; swellings in young apple branches. If these split, disease can enter.
CAUSE Pink-brown aphids secrete the white growth as a defence mechanism, making them impervious to most chemical controls.
REMEDY Hard to control. Prune out worst-affected parts if possible.

APHIDS

PROBLEM Distorted and stunted new growth covered with insects; sticky leaves (excrement); black sooty mould.
CAUSE Plants stressed by lack of water or nutrients are targeted by green, black, or brown sap-sucking insects.
REMEDY Wait for bird and insect predators; wipe off or squash small infestations; cut out affected parts.

APHIDS (BLISTER)

PROBLEM Raised red patches on upper leaf side of currant plants; new growth is puckered or distorted.
CAUSE Clusters of pale yellow sap-sucking aphids on the leaf underside. They rarely cause long-term damage.
REMEDY Encourage predators such as ladybirds and hoverflies with companion planting; squash the aphids.

APPLE SAWFLIES

PROBLEM Young apple fruitlets fall in midsummer and have maggot holes; mature fruits have a ribbon-like scar.
CAUSE Apple sawfly lay eggs in the embryonic fruits; maggots eat fruit.
REMEDY Remove and destroy infected fruits as soon as they're seen in late spring and early summer, before the maggots pupate.

BIG BUD MITES

PROBLEM Blackcurrant buds are overly large and won't open into flowers or fruit; instead, they hold numerous tiny white mites that can spread disease.
CAUSE Mites overwinter in the buds, moving into healthy buds in spring.
REMEDY Remove infested buds on a minimally affected plant. Replace badly infested plants.

GOOSEBERRY SAWFLIES

PROBLEM Rapid defoliation of gooseberry and currant plants.
CAUSE Caterpillar-like larvae of the gooseberry sawfly (pale green with black spots). Up to three generations per year can be present.
REMEDY Inspect regularly; remove and destroy eggs and larvae. Encourage predatory insects and birds.

SLUGS AND SNAILS

PROBLEM Eaten foliage, fruit, and young bark; silvery mucosal trails over plants.
CAUSE Slugs and snails, which will feed on all manner of plants; slugs feed close to the ground; higher damage will be caused by snails.
REMEDY Physical removal of slugs and snails during regular inspections of plants (early morning and late evening).

CODLING MOTHS

PROBLEM Apples, pears, and sometimes quinces are rendered inedible from tunnels and holes in the fruit.
CAUSE Codling moth caterpillars tunnel into the fruit in early to midsummer.
REMEDY Encourage predatory bats and beetles to your garden. Hang sticky pheromone traps in the branches of the trees to catch adults.

RASPBERRY BEETLES

PROBLEM Pale, shrunken patches at ends of raspberries, blackberries, and hybrid berries; small white grubs inside the hole left by plug in fruit after harvesting.
CAUSE Larvae of the raspberry beetle.
REMEDY Wash and inspect fruits carefully before eating. Encourage predatory beetles and birds. Proprietary traps are available.

VINE WEEVILS

PROBLEM Notches eaten from leaves; plants (especially strawberries and containerized plants) wither and die.
CAUSE Adult vine weevils eat the leaves; larvae eat the plants' roots.
REMEDY Remove adults (active at night); encourage predatory beetles and birds; remove affected compost, wash roots, and replant into fresh compost.

FRUIT DISEASES AND DISORDERS

Keeping your plants well watered, mulched, and properly pruned, and practising good hygiene with tools, will help keep your fruit garden healthy and free from disease. Most fruit diseases can be nipped in the bud if they're noticed early enough – simply spending time with your plants will help you spot any signs of infection.

CANKER (BACTERIAL)

PROBLEM On stone fruits (such as plum and cherry), bark sinks inwards and may ooze resin; branches wither and die.
CAUSE A bacteria that is spread during wet, windy weather and enters through wounds to the bark.
REMEDY Prune out affected areas in summer (don't compost at home; clean tools). Replace badly infected trees.

APPLE AND PEAR SCAB

PROBLEM Black or brown growths on skin; cracked, misshapen fruit; leaves with brown or olive-green spots.
CAUSE A fungal infection, prevalent on congested trees and in damp seasons.
REMEDY Remove all fallen leaves (don't compost at home); prune out affected areas of young growth. Thin branches for better air flow.

BROWN ROT

PROBLEM Tree fruits develop soft brown patches with white growths that eventually envelop the fruit.
CAUSE A fungal infection spread by birds, insects, and rain.
REMEDY Remove infected fruits when seen and at harvest time to limit re-infection (don't compost at home). Keep tools clean and disinfected.

CANKER (FUNGAL)

PROBLEM Apple, pear, and mulberry trees show sunken areas of bark that shrink and crack in rings.
CAUSE A fungus that spreads its spores on the wind, entering through wounds and cracks in the bark.
REMEDY Prune out affected areas in summer (don't compost at home; clean tools). Replace badly infected trees.

FROST DAMAGE

PROBLEM Blossom and buds brown and become squidgy; leaves and shoot tips brown, blacken, and can die back.
CAUSE Below-freezing temperatures rupture cells, exacerbated by rapid defrosting, causing extensive damage.
REMEDY Protect blossom and new growth on plants with horticultural fleece, or similar, when frost is forecast.

GREY MOULD

PROBLEM Fruits, especially strawberries and cane fruits, develop a grey mould that envelops the fruit and soon spreads to nearby fruits and young growth.
CAUSE The fungus *Botrytis cinerea*. Spores are present everywhere in the air and infect weak or damaged plants.
REMEDY Remove infected parts, cutting back into healthy growth.

POWDERY MILDEW

PROBLEM A white powder on leaves and stems that can turn yellow and distort; fruits can split; growth is affected.
CAUSE A fungal infection that often affects plants that have been kept in overly dry conditions.
REMEDY Keep plants healthy and well watered, directing water to the soil not foliage; mulch to conserve soil moisture.

FRUIT SPLIT

PROBLEM Maturing fruit, especially pears, apples, currants, and gooseberries, crack and split, browning around splits.
CAUSE Too little and then too much water causes fruit to swell irregularly, ultimately bursting out of its skin.
REMEDY Water plants thoroughly in dry spells; mulch to conserve soil moisture at the surface and aid soil structure.

PEACH LEAF CURL

PROBLEM Peach and nectarine leaves turn pale green then red or purple with a white powder; leaves curl, then fall.
CAUSE A fungus spread by wind and water splash.
REMEDY Fruit and replacement leaves are usually unaffected. Remove affected parts when seen; protect outdoor fruit from rain from midwinter to spring.

SILVERLEAF

PROBLEM Stone fruit trees, especially cherry and plum, have a silvery sheen to their leaves, affecting growth in the long term.
CAUSE A fungus that infects trees through pruning cuts or insect damage.
REMEDY Prune stone fruits in summer when infection is least likely, removing affected parts and disinfecting tools.

WORKING WITH NATURE

Working with nature, rather than waging a constant war against it, is a far more relaxing, enjoyable, and productive way to garden. A healthy, balanced, and biodiverse garden ecosystem will benefit both you and the environment. Enlist the help and support of natural predators and beneficial insects and creatures so that everyone can enjoy the bounty of your fruit garden.

Providing cool, shady hiding spaces for toads and frogs will encourage them to your garden.

WELCOMING HELPFUL WILDLIFE

Adding fruit plants to your garden will attract more wildlife, but including helpful habitats will encourage them to stay. A pond is especially useful, no matter how small, but frogs and toads don't need water if there are some damp, shady hiding holes at the back of borders or in corners, such as terracotta pots laid on their sides. They will also enjoy log piles, as will beetles.

Other insects prefer dry places: add bug hotels on sunny walls around your plot (a few small ones are better than one large hotel). An easy way to create one is to fix lots of short lengths of bamboo cane and other stems and twigs into a box or pot, or drill holes into a block of wood.

Bird feeders and tables will attract birds to your garden; add nesting boxes, if you have space, so that they raise their young on a diet of aphids and caterpillars from your garden.

> **TOP TIP** CLEAN YOUR BIRD TABLES AND BIRD FEEDERS WITH HOT, SOAPY WATER TO PREVENT DROPPINGS, MOULD, BACTERIA, AND DISEASE FROM GATHERING. WASH YOUR HANDS AFTER CLEANING.

This garden's bug hotel and spring flowers will help attract beneficial insects.

NATURAL PEST CONTROL

The table below shows primary fruit pests and their natural predators. Birds are of huge benefit to fruit gardeners but can also be a pest on crops; keep bird feeders topped up and they're less likely to go for your fruit (see p.44).

Ladybird larvae eat numerous aphids and are voracious pest predators as adults.

PEST	NATURAL PREDATORS
Aphids	Ladybirds and their larvae; lacewing larvae; hoverfly larvae; beetles and other insects; blue tits and other birds
Caterpillars	Birds such as blue tits, great tits, sparrows, and blackbirds; beetles; hedgehogs; wasps
Moths (e.g. codling moth)	Bats
Slugs and snails	Toads, frogs, and newts; birds such as thrushes; beetles; hedgehogs

REDUCING WEEDS

Bare soil isn't natural and will soon be colonized by unwanted plants: namely, weeds. By planting densely around your fruit trees and bushes, leaving little to no bare soil, you can reduce the chances that weeds have to germinate and spread – and it looks prettier too.

Plant herbs, bulbs, or ornamental flowering plants around the base of fruit trees and bushes in beds and borders so that there will be no empty space (see pp.16–17). Avoid sowing vegetable or annual flower seeds, as you don't want to disturb the roots of fruit plants when digging up the old plants in autumn. You can also plant around the base of fruit trees in pots; herbs such as oregano, mint, and thyme are ideal for this.

The dense planting of herbs under this apple tree will reduce weeding and provide tasty harvests.

PLANTING FOR POLLINATORS

Without pollination of fruit blossom there will be no fruit to harvest. The more flowers you have in your garden the better, but some blossom (such as apple, pear, cherry, and quince) is more obvious to pollinating insects than others (blackcurrant and gooseberry, for example). Planting flowers that bloom at the same time in spring as your fruit crops will help to attract bees and other pollinators.

The daffodils planted under this cherry tree will help attract pollinators.

NEED TO KNOW
These spring flowers will provide a feast of nectar and pollen to attract pollinators:
- Bulbs such as snowdrops, tulips, and crocuses.
- Perennials such as hellebores, primroses, rosemary, wood anemones, and pulmonaria.
- Biennials such as wallflowers, honesty, and foxgloves
- Lawn wildflowers such as dandelions and self-heal.

PROTECTING YOUR CROPS

Cultivating a biodiverse garden ecosystem will certainly help to limit problems with pests and diseases attacking your fruit, but it may sometimes be necessary to put additional measures in place. If you're keen to protect your harvests from damage, there are various ways to achieve this without risking harm to wildlife such as birds, hedgehogs, frogs, and bees.

Squirrels will love your fruit – try to deter them by being in your garden as often as possible.

ARE THEY PESTS?

It's always worth considering whether the potential pest is actually a problem. Birds are helpful and entertaining in the garden, and it might be worth tolerating the loss of a few berries in return for their caterpillar-eating, antics, and song. Aphids cause minimal long-term damage; after a short lag their presence will attract predators such as ladybirds, for whom they're essential food. Being patient may be better than the alternative of spraying your fruit with chemicals.

Hang shiny objects in tree branches to deter birds as the fruit begins to ripen; remove them after the harvest.

DETERRING BIRDS

Birds are attracted to red berries, so consider planting fruit of other colours, such as white strawberries and currants. Keep bird feeders topped up but placed away from your fruit plants. Add bird scarers to fruit trees and bushes, such as scarecrows or shiny objects that glint in sunlight, or buy fake predators (kites in a bird silhouette on a long, bendy pole). Your mere presence in the garden will deter birds, albeit temporarily.

A great tit is more likely to eat aphids and caterpillars than the gooseberries on this bush.

WILDLIFE-FRIENDLY PROTECTION

Birds, amphibians, reptiles, and small mammals can all become ensnared by netting used to cover edible crops, and often die as a result. If you consider you have no alternative but to put up physical barriers between birds (and other pests) and your crops, use horticultural fleece or fine mesh instead. These are available as bags for small trees and bushes, and as large sheets that can be stretched over metal hoops to make a tunnel, or wrapped around plants and tied in place. Ensure that the protection is securely fastened all the way around the edge so wildlife won't get trapped. A longer-term investment is a permanent fruit cage: but, again, ensure the walls and roof are secured and wildlife-friendly. If rabbits or deer are likely to be a problem, you'll need to protect the trunks of trees with guards. For pests such as slugs and vine weevil (see p.39), physically removing them yourself is the most environmentally friendly option.

A fine mesh is preferable to netting but still needs to be securely fastened around the crop to avoid harming wildlife.

NEED TO KNOW

- Use chemical sprays against pests only as a last resort as they also kill beneficial insects, such as bees.
- Insect populations, crucial to our existence on this planet, are suffering a massive global collapse.
- Many sprays have been withdrawn from home use. If you choose to use sprays, check they're suitable for edible crops and follow the label instructions closely.
- Homemade sprays against aphids, using infusions of garlic or chilli peppers, can be effective, but their strength is impossible to quantify and their long-term effects on the plants and environment are unknown.

Protect flowering fruit trees and bushes with horticultural fleece when cold weather is forecast.

PROTECTING BLOSSOM

Late frosts can damage blossom, meaning it never develop into fruit. For larger and freestanding trees this often means you just have to chalk it up to a bad year, but smaller, pot-grown, and wall-trained trees can be shielded from the worst of the cold more easily. Move pots under cover, if possible: an awning or patio umbrella will often be sufficient. Use horticultural fleece sheets or bags to cover plants you can't move: try to make sure the fleece isn't touching the blossom. If late frost is a problem in your area, it can be worth constructing a metal or wooden framework around wall-trained fruit, over which plastic or fleece sheeting can be easily secured. Cover plants for the night only, so that pollinators have access in the day.

THE HARVEST

Picking and savouring your homegrown fruit is the highlight of the growing year. With a little space and planning, you can feast on your own produce from spring to autumn, and have enough to preserve to enjoy in winter as well.

When the harvest is over, it's time to sort out the garden for winter and plan ahead. Think about which varieties you planted, how they performed, and which ones you enjoyed most – this will help you decide what to grow again next year and which new fruit to try.

STORING AND PRESERVING

Some fruits such as apples and quinces can be stored without any treatment in a cold place for several weeks or even months (see *individual entries, pp.52–135, for storage information, if applicable to the fruit*). Other surpluses need to be cooked or otherwise preserved so that your harvest isn't wasted.

Freezing raw is ideal for fruits such as berries, currants, plums, kiwis, and cherries. Freeze them either so you can use them through the winter, or just to store them temporarily until you have time to make them into something else such as jam or ice-cream. Wash and dry the fruit first, remove any part you don't want to eat (stalks, stones, or skin), and chop into pieces if necessary. To avoid the fruit freezing together in a solid block, freeze it first in a single layer on baking trays before transferring to boxes or bags after 24 hours. Other fruits can be roasted or stewed then frozen in tubs.

The other ways to preserve fruit are in sugar – jams, jellies, compotes, fruit leathers (dried fruit), or bottled in syrup – or in alcohol or vinegar. Fruit vinegars involve steeping the fruit in vinegar; flavoured spirits such as damson gin are similarly infused. If you're feeling more ambitious you could try making your own wine, either from grapes or (like the old country wines) strawberries, cherries, or blackberries.

> **TOP TIP** KEEP A JOURNAL OF YOUR ACTIVITY IN THE GARDEN TO REFER BACK TO IN FUTURE. JOT DOWN WHAT YOU PLANTED, HOW WELL IT GREW, AND IF YOU ENCOUNTERED ANY PROBLEMS.

Give your tools a good clean at the end of the season so you're ready for the year ahead.

AFTER THE HARVEST

Once the fruit has been preserved or eaten, it's time to organize the garden for winter. Anything that might harbour plant diseases should be removed, such as fallen fruit with brown rot or infected leaves. Make leaf mould with leaves and turn the compost heap (see p.27), but leave some piles of leaves to give wildlife a place to nestle in winter.

Move any plants that need protecting from the cold and wet to their winter homes, and give extra protection to those still outside, such as wall-trained figs. Check the ties on any stakes and trained fruit wires, loosening or replacing them as appropriate. Secure loose canes or stems of climbers that might get broken in winter winds. Finally, clean your tools and equipment so that they're ready for winter pruning later in the year.

Wrapping fruits in newspaper before storing on trays stops them touching and helps prevent bruising and rotting.

Making jam from your own fruit means you can have a taste of your summer garden in the depths of winter.

If you have a glut of fruit that you don't want, give it away
to friends, family, or the local food bank, or sell it from your gate.

EXPANDING YOUR FRUIT GARDEN

As fruit trees and bushes are perennials, you don't need to get fresh plants for next year – unless you want to grow more. For most fruits, it's far easier to buy them, but you can double or triple your stock of strawberry plants easily and for free by rooting runners. Sowing fruit pips is longer-term investment, as they take a long time to mature from seed – but it's still a fun experiment.

Grow your cherry stones into new trees and then name the fruit yourself.

> **TOP TIP** THE BEST TIME TO SOW FRUIT SEEDS OF HARDY TREES IS AUTUMN, AS THEY NEED A COLD SPELL OUTSIDE TO STIMULATE THEM INTO GROWTH. CITRUS SHOULD BE SOWN IN SPRING.

This plantlet has been secured, and now the remaining loose end of the runner needs to be removed.

STRAWBERRY PLANTS FOR FREE

Every summer, strawberry plants produce "runners" – long stems with baby plantlets growing on them that spread out from the main ("mother") plant. Peg down the first plantlet on one or two runners into a small pot of multipurpose compost. Water the pots so the compost remains damp. Keep the plantlet attached to the mother plant by the runner until the plantlets develop roots, then cut off the runner stem (see above). If you don't want all the new plants, give them to friends and family.

SAVING SEED

Most fruit species won't "come true", meaning that the variety of the fruit on the seed-grown plant won't be the same as that of the fruit it came from. Unfortunately, the trees are often far too vigorous (this is because they aren't controlled by a rootstock in the same way as professionally grown trees) and the fruit is also inferior. However, it can be great fun to sow pips or stones from your fruit and then discover what interesting results emerge.

If you grow more than one species of citrus, the seedling fruits could be a cross of those species.

CREATING A FAMILY TREE

Commercial growers use a process called grafting to increase their stock of fruit trees, whereby a bud from the variety species is cut away and strapped onto a wounded spot of the rootstock stem (see pp.245). The two then fuse together and grow as one tree.

You can create your own space-saving "family tree" quite simply by grafting buds of different varieties of apple onto separate branches of one young tree (see opposite and below).

The bud from the variety tree is secured onto the branch with tape until the two have fused.

Every branch of this ancient apple grows a different variety, each one labelled to avoid confusion.

There are lots of plum varieties to choose from, each with different tasty attributes, and a mature tree can produce enough fruit to keep you well stocked for the rest of the year.

FRUIT TREES

Fruit trees are among the most romantic of trees. Who can resist wandering among their spring blooms; sitting beneath a cherry tree in summer, plucking sun-warmed fruits off the branches; or biting into a juicy apple on a crisp autumn day?

TERRIFIC CHOICE

Growing your own fruit trees gives you a far greater choice of fruits, and varieties of those fruits, than is available to buy in shops. Crops such as medlars, quinces, mulberries, and damsons are rarely seen for sale, and yet they're delicious and easy to grow. Apples, plums, and cherries might be readily obtainable but you can grow many more varieties than the few sold in the supermarkets. Fresh homegrown fruit in season is more nutritious and far tastier than anything you can buy.

COMPACT STRUCTURE

You don't need a large garden to grow fruit trees – you can grow them on a balcony or your front doorstep. Apples, pears, cherries, and plums can all be easily trained into compact forms that will grow well in pots or onto walls and fences, and they'll still provide a great harvest. Figs and persimmons benefit from being grown in pots; and being able to move potted citrus, peaches, and apricots under cover for winter helps them thrive.

However large the trees you grow, they'll add structure to the garden and can be enjoyed all year. Plum and cherry trees in particular have brilliant autumn colours, and a garden full of blossom in spring brings cheer after the starkness of winter.

WHERE TO GROW

Fruit trees prefer moist, well-drained soil but will grow in most soils, provided they're not prone to waterlogging over winter. Trees in free-draining, sandy soils will need more watering while they're in growth from spring to summer. Ideally, give your trees a sunny, warm, and sheltered spot – they'll fruit best there – especially for the trees native to Mediterranean climates, such as peaches, apricots, citrus, and persimmons . Hardier apples, pears, plums, and cherries will grow in most situations, and acid cherries can even grow well on north-facing walls. Training fruit onto walls (especially brick) helps them grow and the fruit ripen: the wall absorbs the sun's heat in the day and releases it slowly overnight, keeping the ambient temperature around the tree warmer than the rest of the garden.

CITRUS *CITRUS* SPECIES

The citrus family includes lemons, oranges, and limes as well as oddities such as 'Fingered', and aromatics such as bergamot. Citrus blossom has a glorious scent: grow the trees as houseplants in winter and keep in a nearby sunny spot outdoors in summer to enjoy the aroma and the fruit.

GROW FROM Potted trees
HEIGHT AND SPREAD Up to 6 × 2.5m (20 × 8ft)
HARDINESS Tender
IDEAL SOIL Moist but well-drained
POSITION/SITE REQUIREMENTS ☼
TIME TO MATURITY Three years

CALENDAR

	SPRING	SUMMER	AUTUMN	WINTER
PLANT				
HARVEST				

Through the year
Although their main season is late winter, citrus can flower and fruit at any time of year and often at the same time.

Regular misting will help keep the air around citrus trees humid and will also assist with pollination under cover.

Place citrus trees in a sunny, warm, and sheltered spot outdoors for summer.

PLANT

In temperate climates, citrus need moving under cover for the winter (or growing there all year round), so they're best planted in a container that's easily moved. Most plants are sold as grafted specimens, so choose a dwarfing rootstock to help keep your tree small enough to grow indoors (see *p.24*). Indoors, citrus need high humidity and lots of bright light – keep the trees away from radiators and other heat sources. Outdoors, although technically they're hardy to 0°C (32°F), they won't survive that temperature for long and need to be moved under cover before nighttime temperatures drop below 10–14°C (50–57°F).

IN CONTAINERS Plant into a large container using a peat-free, soil-based compost. Repot into a larger container annually until the pot is around 50cm (20in) in diameter, and then continue to repot annually with fresh compost once the tree has reached its ultimate size. Depending on their size, trees may need staking with a strong cane.

GROW

Keep citrus trees well watered and give them a liquid feed monthly in spring and summer. In autumn and winter let the compost almost dry out before watering thoroughly. The warmer and sunnier the place you keep your citrus tree the better the fruit quality will be, and the more fruits you'll get. Aim to keep them no colder than 14°C (57°F) at all times, and at around 20–25°C (68–77°F) daytime temperature in summer. Trees left in the cold or a draught will probably lose their fruit. If there's insufficient light over winter they may lose their leaves, but they'll regrow in spring. Prune trees as required in early spring to shape them and keep them to size: take care as the branches are spiny. Remove any shoots growing up around the base of the stem.

'Fingered' or 'Buddha's Hand' citrus is named for its unusual shape.

'Nagami' kumquats have a sweet skin and sour flesh that are best eaten together.

VARIETIES

Choose a citrus tree depending on what type (species) of fruit you want to grow, such as lemons, grapefruit, or calamondin, which itself will depend on how you use citrus fruits – for drinks, cooking, or preserves. A few fruits have different varieties from which to choose, others are just different species. The botanical names are extremely varied, so the full name for each fruit is given in brackets here (*see right*) to avoid any confusion. All are self-pollinating.

CITRUS FINGERED (*Citrus medica* 'Fingered', syn. *C.* 'Buddha's Hand') The large yellow citrus fruits consist mostly of pith, but the zest and rind are extremely fragrant.

GRAPEFRUIT STAR RUBY (*C. × aurantium* Grapefruit Group 'Star Ruby') Moderately vigorous trees produce sweet fruits that have a deep red flesh and thin skin.

KUMQUAT NAGAMI (*C. japonica* 'Nagami') Kumquats are the hardiest citrus plants. Oval orange fruits (eaten whole) hang in prolific clusters.

LEMON GARY'S EUREKA (*C. × limon* 'Gary's Eureka') A prolific, relatively

hardy lemon with juicy, seedless fruits. MEYER (*C. × meyeri*) Excellent quality fruit, much valued in culinary circles, is borne on compact plants that are relatively tolerant of cool conditions.

LIME TAHITI (*C. × aurantiifolia* 'Tahiti', syn. 'Persian') Compact, productive trees bear relatively sweet fruit with a pale green skin.

ORANGE CALAMONDIN (*C. × macrocarpa*) Tiny fruits, each less than 5cm (2in) across, are good for preserves. Needs the warmest temperatures of all citrus. WASHINGTON (*C. sinensis* 'Washington'). A juicy and seedless navel orange that ripens in late autumn. Trees are vigorous.

HARVEST

Fruits are ripe when the skin turns a rich colour. Pick by snipping off the branch with secateurs. Once ripe, they can be left on the tree for some time or stored in the fridge for weeks. Expect each tree to yield around 10–12 fruits per year. If you have enough to preserve, freeze sliced fruit to add to drinks, or make marmalade or preserved lemons.

Citrus trees, such as this lemon, often flower and fruit at the same time.

QUINCE *CYDONIA OBLONGA*

Quinces have pretty blossom and leaves, but their perfumed fruits are the highlight. In temperate climates they rarely ripen sufficiently to eat raw, but they're sublime when cooked. Try making quince chutney or membrillo (quince jelly) for your cheeseboard, or add quince to an apple pie.

GROW FROM Potted or bare-root trees
HEIGHT AND SPREAD Up to 4.5 × 4.5m (15 ×15ft)
HARDINESS Hardy
IDEAL SOIL Moist but well-drained
POSITION/SITE REQUIREMENTS ☼
TIME TO MATURITY Two to four years

CALENDAR

	SPRING	SUMMER	AUTUMN	WINTER
PLANT			▨	▨
HARVEST			▨	

Through the year Quince fruit can continue ripening on the tree even after the leaves have fallen in autumn.

A downy fluff covers quince fruits as they grow: wash this off under running water before cooking.

PLANT

Plant quince trees in a sheltered, sunny spot, avoiding frost pockets, and at least 4m (12ft) from other trees. There are two rootstock choices: varieties grown on a Quince A rootstock will reach 3.5–4.5m (11–15ft), while those on a Quince C rootstock will grow to 3–3.5m (10–11ft) (see p.24).

Plant during the dormant season, avoiding frozen ground, and water and mulch the tree afterwards with compost or other well-rotted organic matter. Support the tree with a stake for the first two years, checking the ties regularly and loosening them as required (see p.28).

IN CONTAINERS Trees on a Quince A rootstock are suitable for larger raised beds; for large containers, use varieties sold as "dwarf" or "patio" quince (1.5–2m/5–6ft 6in in height and spread). Use a peat-free, soil-based compost.

Water quince trees well after planting and allow each application to soak in before adding more.

GROW

Establish a good framework with formative pruning (see p.82) as quince trees have a naturally irregular habit. After this, they'll need only minimal pruning every winter (see p.84).

Water young trees in the ground during dry spells. Container-grown trees should be watered regularly to avoid the fruit splitting as it swells; give them a liquid feed fortnightly in spring and summer. Mulch trees annually with compost or other well-rotted organic matter mixed with a handful of wood ash. Repot trees in containers every other year.

Quince can be troubled by powdery mildew and brown rot (see pp.40–41). Burn fallen leaves and pruned shoots from infected trees.

Some quince varieties are pear-shaped, as here; others are rounder and apple-shaped (*see below*).

VARIETIES

All quince trees are self-pollinating. Japanese quince (*Chaenomeles*) is a thorny spring-flowering shrub and not true quince, although its small fruits are edible. There's not much divergence between quince varieties in terms of taste, but their size and shape differ.

AROMATNAYA Smooth, rounded shape; fruit can be eaten fresh if a warm autumn allows sufficient ripening.

CHAMPION A reliable crop of pear-shaped fruits; tree has good disease resistance.

MEECH'S PROLIFIC A heritage variety (*see p.22*) with strongly scented, large, pear-shaped fruits and pink-tinged blossom.

PORTUGAL (syn. 'Lusitanica') Somewhat lumpy, pear-shaped fruits. Flesh turns pink when cooked; the trees are vigorous but less hardy.

SERBIAN GOLD Reliable crops of apple-shaped fruits. Award of Garden Merit (AGM, *see p.23*).

VRANJA NENADOVIC Pear-shaped, aromatic fruits; ready to pick in early autumn.

HARVEST

Quinces are ready when they're golden, aromatic, and lift easily off the branch; in a long, warm autumn they may ripen partially or completely on the tree. Pick before the first frosts. Leave the fruit to ripen and mellow further indoors for six to eight weeks. Even when ripe the flesh can be pithy, and quinces are often used only for cooking and preserving. Don't store or display quinces with other fruits such as apples and pears, as their intense scent will infuse them. If stored in a cool place, quinces will keep for several months.

Yields are variable as the blossom is susceptible to frost, but a fully grown tree can produce 15kg (33lb) of fruit.

Keep quinces in a bowl on their own so their perfume doesn't flavour other fruit.

PERSIMMON *DIOSPYROS KAKI*

The persimmon's tropical flavour and glorious fluorescent colour belie the fact that it's easily grown in temperate climates. Recent advances in breeding are reducing the time it takes for a persimmon tree to bear its first fruit – but whatever the wait, it's definitely worth it.

GROW FROM Potted or bare-root trees
HEIGHT AND SPREAD Up to 8 × 4m (26 × 13ft)
HARDINESS Hardy
IDEAL SOIL Moist but well-drained
POSITION/SITE REQUIREMENTS ☼
TIME TO MATURITY 5–10 years

CALENDAR

	SPRING	SUMMER	AUTUMN	WINTER
PLANT				
HARVEST				

Through the year
With its vibrant autumn foliage and fruit until early winter, persimmon brings great seasonal interest to the garden.

Thin the fruit to a maximum of two per shoot to prevent breaking branches and an irregular crop load (see *pp.84–85*).

GROW

Persimmons need as long a growing season as possible: get them into growth quickly in spring by bringing under cover or wrapping the tree or pot in fleece to warm it. Keep sheltered all year.

Establish a good framework with formative pruning (see *pp.82–83*). After this, they'll only need minimal pruning every summer (see *pp.84–85*). Snip off any suckers (shoots coming up from the base of the trunk). Keep the compost of potted trees moist but not sodden. Give a liquid feed fortnightly in spring and early summer – feeding after that can risk frost damage to new growth.

An annual, dark-coloured mulch in late winter – compost and biochar, for example (see *p.26*) – will help warm the roots. Repot mature trees in containers every other year.

Persimmons can be allowed to form a spreading tree, or pruned to keep to size.

PLANT

Persimmons (also known as Sharon fruits or kakis) like a sunny, sheltered spot, with space for their spreading branches. If you buy a tree on a *Diospyros virginiana* rootstock (see *p.24*) it will give more vigour and hardiness and bring forward the fruiting date. Grow in a container – repotting annually into a larger pot – until the tree starts fruiting. Plant out in the dormant season: stake, water, and mulch the tree afterwards (see *p.26*).

IN CONTAINERS Persimmons can be kept in containers. Use a container of 50cm (20in) in diameter for a fully grown tree, filled with peat-free, soil-based compost. Move potted trees under cover for winter, if possible.

VARIETIES

All persimmons are self-pollinating and are divided into astringent and non-astringent varieties (astringency is the mouth-puckering dryness caused by tannin compounds). The astringent varieties are generally eaten only after they've ripened – during the ripening process, the astringency disappears and the flesh turns sweeter and more juicy.

Recent modern breeding techniques have produced non-astringent varieties with fewer bitter tannins – these can be eaten when golden-coloured but still crunchy (unripe). However, these persimmons can be less flavourful and less hardy than the astringent ones. In temperate climates, even non-astringent persimmons will retain their tannins until they're fully ripe, unless the summer has been particularly warm and long.

There are hardier, non-astringent varieties, such as 'Fuyu', but in colder areas it's preferable to choose an easier-to-grow astringent variety and ripen the fruit after harvest by misting with alcohol (see right).

CHOCOLATE Astringent. Small fruit with bright red skin. Grow alongside another variety for better pollination, after which the flesh turns dark-coloured and richly flavoured with a hint of spice.

FUYU Non-astringent but has good cold-tolerance. Popular commercial variety.

GIANT FUYU This variety has the same properties as 'Fuyu', but its fruits can be as much as 40 per cent bigger.

IZU Non-astringent. Dwarf variety that ripens slightly earlier. Good choice for temperate climates.

ROJO BRILLLIANTE Astringent. Red skin and rich flavour. Widely grown commercially.

Cut, don't pull, the fruit stalk as the brittle branches are easily damaged.

HARVEST

Pick before the first frost; the fruits should be a deep orange colour. They're unlikely to ripen naturally on the tree in the UK climate and will probably still be hard, but they'll continue to ripen indoors. This takes one or two weeks in a fruit bowl in a warm room. Eat when still crunchy or once ripened to a squashy texture.

To ripen them in around five days, mist the fruit with a high-percentage alcohol, such as vodka, then seal in a plastic bag with a banana or apple. If kept at room temperature, the flesh of even the most astringent varieties will turn into a sweet, smooth jelly.

A mature tree can yield around 22kg (49lb) of fruit – potentially more, depending on the size of the tree.

Persimmon trees hold their golden fruit on branches into midwinter, but you should harvest the fruit before the first frosts.

FIG *FICUS CARICA*

Sweet and sticky, a fresh fig plucked straight from the tree is a wonderful late-summer treat. The plant's aromatic leaves can be used too, for infusing into creamy desserts and sorbets. Figs are ideal for growing in courtyards or against a wall, where they make the most of the sun's warmth.

GROW FROM Potted plants
HEIGHT AND SPREAD Up to 3.5 × 2.5m (11 × 8ft)
HARDINESS Hardy
IDEAL SOIL Moist but well-drained
POSITION/SITE REQUIREMENTS ☼
TIME TO MATURITY Two to three years

CALENDAR

	SPRING	SUMMER	AUTUMN	WINTER
PLANT				
HARVEST				

Through the year
Figs are deciduous, with attractive foliage; their large leathery green leaves turn bright yellow in autumn.

PLANT

Figs are best trained against a south- or southwest-facing wall. Restricting the roots puts the tree's energy into fruiting (not vigorous foliage growth) and keeps the plant compact. If growing against a wall, plant into a pot, a raised bed over a solid base, a narrow border, or a planting pit. Fix the training wires before planting in spring (see p.86).

To make a fig pit, dig a 45cm- (18in-) deep pit into the soil at the base of a wall (see right) and line with three 60cm- (24in-) square paving slabs. The slabs should protrude 2.5cm (1in) above the soil/paving level to stop the roots from running over the surface. Put a layer of rubble in the base, then plant into a mix of topsoil and gravel in a ratio of 3:1.

IN CONTAINERS Figs will grow well in a large container filled with a soil-based, peat-free compost. Every other year remove freestanding plants from their pot, knock off any loose compost and cut back any thick roots, then replant with fresh compost, back into either the same pot or a slightly larger one.

Walls absorb heat from the sun then release it at night, helping to ripen the figs.

Figs make a great houseplant for a warm, sunny spot.

A planting pit for wall-trained figs is simple to construct (see left).

Pea-sized baby figs develop in the join between the leaf and stem.

GROW

Protect figs from frost by moving them into a greenhouse or indoors for winter, or by covering them with horticultural fleece. Remove the coverings or put pots back outdoors in mid-spring.

Prune figs in late spring and summer to keep to shape and to promote fruiting (see pp.84–85); take care when pruning as the sap can cause allergic skin reactions in some people.

The fruit develops in two stages: first, baby figs overwinter on the plant; then, provided they're not damaged by frost, these swell and ripen to be ready to harvest the following midsummer. Remove any larger fruits that didn't ripen at the end of autumn.

Water well for the first two years; thereafter, water in dry spells. Figs in containers should be watered regularly, to prevent the compost from drying out, and given a liquid feed fortnightly during the spring and summer.

Give your fig trees an annual mulch of compost or other well-rotted organic matter mixed with a handful of wood ash.

HARVEST

Figs are ready to harvest from midsummer through to early autumn, or even beyond. The fruit is ripe when it's fully coloured, hanging down, soft, and with splits appearing around the stalk. Pick the fruit by holding the stalk and breaking it from the stem. Figs don't keep for long after picking, so either eat them fresh or preserve them. In a good year, a mature fig tree will yield around 15kg (33lb) of fruit.

A deeply coloured ripe fig is surrounded by fruits of next year's crop.

Ripening 'Brunswick' figs are paler in colour than some other varieties.

VARIETIES

Fig fruits develop without the need for pollination. In cool climates they are unlikely to be pollinated at all and will develop without seeds.

BOURJASSOTTE GRISE Late season. Blue-skinned fruits with a rich-tasting brown flesh. Good for containers under glass, unlikely to ripen outside in a temperate climate.

BROWN TURKEY Mid-season. A reliable cropper in temperate climates; Award of Garden Merit (AGM, see p.23).

BRUNSWICK Mid-season. Green- or brown-skinned fruits, with an excellent rose-scented flavour; reliably crops in cool climates.

EXCEL Mid- to late season. Green- or yellow-skinned fruits with honey-scented amber flesh. Vigorous and tolerant of cold weather.

LITTLE MISS FIGGY Mid-season. A dwarf cultivar bred from 'Rouge de Bordeaux' (see below); ideal for containers.

OSBORN'S PROLIFIC Mid- to late season. Sweet fruit with a jammy texture. Fruit ripens well outside in colder areas.

ROUGE DE BORDEAUX (syn. 'Petite Negri' or 'Violette de Bordeaux') Mid-season. Small, purple-skinned fruit, with jammy, dark red flesh. Good for containers, indoors or out.

WHITE MARSEILLE Early season. Large, yellow- or green-skinned fruit with smooth, very sweet flesh. Grown in the UK since the 1500s.

APPLE *MALUS DOMESTICA*

From nutty to floral, sharp to sweet, and crunchy to soft, there's an apple for everyone. And with apple trees bred for the smallest spaces, you can have a mini-orchard on your doorstep or balcony. Visit an apple day or autumn garden show to taste different varieties before deciding what to buy.

GROW FROM Potted or bare-root trees
HEIGHT AND SPREAD Up to 4 × 8m (13 × 26ft)
HARDINESS Hardy
IDEAL SOIL Moist but well-drained
POSITION/SITE REQUIREMENTS ☼ ☼
TIME TO MATURITY Three years or more

CALENDAR

	SPRING	SUMMER	AUTUMN	WINTER
PLANT				▨▨
HARVEST			▨▨	

Through the year
After the blossom fades, watch the apples swell on the branch through summer in anticipation of the autumn harvest.

Container-grown trees can yield a good harvest for the limited space they take up.

PLANT

The right choice of rootstock (see *pp.24–25*) will ensure your apple trees don't get out of hand and will be productive in the space you can allow them. Freestanding trees on an M25 rootstock need spacing at least 6m (20ft) from other trees, but cordons grown on an M27 rootstock can be as little as 75cm (30in) apart. When training apples as cordons, stepovers, or espaliers (see *pp.86–87*), put the training wires and the posts in before planting.

Apples flower and fruit best in a sunny spot, but they'll also take some dappled shade. Cooking apples will even grow well on a north-facing wall, allowing you to utilize the shadiest spots of your garden for a harvest of fruit.

Plant bare-root and potted apple trees in the dormant season, ideally in autumn while the soil is still warm, but any time between autumn and late winter. After planting, water the tree well then apply a mulch of compost or other well-rotted organic matter around the base of the trunk.

Planting a line of cordons at a 45° angle adds visual interest and makes them more productive.

IN CONTAINERS The best apple trees for containers are those sold specifically for growing in pots, often as "patio" or "ballerina" trees, which are on dwarf rootstocks and should be pruned as cordons. Otherwise, any tree on an M27 or M26 rootstock will grow in a deep pot that's at least 50cm (20in) in diameter, and can be treated as a cordon or a goblet tree (see *p.86 and p.84*). Plant into a soil-based, peat-free compost and water well afterwards. Drive a long, thick bamboo cane or other long pole (such as a hazel bean pole) right to the base of the pot, close to the tree, and tie it in for support. For planting in raised beds, allow a depth of at least 45cm (18in) and proceed as for containers.

Gather and dispose of any fallen fruit with brown rot to reduce the chance of it infecting healthy apples.

HARVEST

A ripe apple detaches easily from the stalk when lifted slightly and given a quarter turn. Later-season apples can be picked a little early and ripened and stored indoors if frosts or storms that would blow them to the ground have been forecast.

Apples store well in a cold place, with slight humidity to prevent them from shrivelling. Lay undamaged fruits in trays in a single layer, not touching; later-season apples store better than early season varieties. Check them regularly and remove any rotten fruit. Alternatively, preserve them as juice or cook and then freeze them.

Yields are dependent on the size of the tree and variety as well as the growing conditions – for example, a late frost can kill the blossom, reducing the harvest. A single cordon could bear 2.25–4.5kg (5–10lb), an espalier 13.5–18kg (30–40lb), and a large mature freestanding tree between 27–55kg (60–121lb).

Moulded cardboard liners are ideal for preventing apples in storage from touching. Ask your grocer for spare ones.

GROW

Water young trees regularly; water mature trees in dry spells, especially during flowering and fruiting. Water trees in pots regularly (they may need it daily), ensuring the compost doesn't dry out as irregular watering can lead to wonky or split fruit. Give trees in containers a liquid feed every fortnight during spring and summer. Mulch trees annually in late winter with compost or other well-rotted organic matter.

Apple trees will need first formative and then maintenance pruning every winter, while the trees are dormant (see p.82 and p.84). Trained forms will also need summer pruning (see pp.88–89). Apples can be affected by canker, apple scab, and brown rot (see p.40); aphids and codling moth can be a problem too (see pp.38–39).

'Egremont Russet' has the distinctive mottled, quite rough skin of russet apples.

'Jonagold' has attractive golden-green skin striped with orange and red.

A "stepover" or low-growing tree gives a productive edge to a border – as here, with the red-skinned variety 'Falstaff'.

VARIETIES

There are thousands of apple varieties to choose from, so whittle them down to a shortlist based on taste and then check their pollination group (PG) – which indicates when they'll flower – and if they're available on the rootstock you need (see p.60 and p.24). "Russet" refers to the rough skins that are a trait of some apple varieties – don't be put off by this, as they have some of the best flavours. In the lists below, "PG" denotes the pollination group number.

DESSERT/EATING APPLES

ASHMEAD'S KERNEL Mid- to late season. PG4. Heritage russet apple (see p.23) with a delectable flavour. Award of Garden Merit (AGM, see p.23).

BEAUTY OF BATH Early season. PG2. This heritage variety has fragrant, slightly tart, red-streaked flesh. Can't be stored but makes good juice.

COX'S ORANGE PIPPIN Early to mid-season. PG3. Aromatic, crisp, and juicy heritage variety, a superb eating apple. Buy the self-pollinating version ('Cox's Self-fertile') for better yields.

EGREMONT RUSSET Mid-season. PG2. The crisp, juicy flesh has a nutty flavour. A heritage variety suited to northern areas. AGM.

ELLISON'S ORANGE Early to mid-season. PG4. Fruits have an aniseed aroma and attractive red-flushed skin. Trees have some resistance to frost, scab, and canker. AGM.

FALSTAFF Mid- to late season. PG3. Juicy, flavourful apple that was the result of cross-breeding 'Golden Delicious' and 'James Grieve' in the 1970s.

GREENSLEEVES Early to mid-season. PG3. This variety gives reliable crops of pale-skinned, juicy fruit with a good acid/sweet balance. AGM.

JONAGOLD Mid-season. PG3. Reliable and vigorous variety that stores well; juicy flesh has a honeyed flavour. AGM.

KIDD'S ORANGE RED Early to mid-season. PG3. Rich, aromatic flavour similar to 'Cox's Orange Pippin' (see left). AGM.

ORLEANS REINETTE Mid- to late season. PG4. This heritage variety bears large apples of excellent flavour and fine texture.

PITMASTON PINE APPLE Mid- to late season. PG3. Small yellow-skinned fruits have a nutty flavour and distinct honeyed pineapple scent.

SCRUMPTIOUS Mid-season. PG3. Thin, red skins and crisp, sweet flesh; these are reliable trees that crop and hold the fruit well. AGM.

COOKING APPLES

ANNIE ELIZABETH Mid- to late season. PG4. Good crops from disease-resistant trees. Light-flavoured heritage apple that keeps its shape when cooked.

BLENHEIM ORANGE Mid- to late season. PG3. Heritage variety; dual-purpose fruits can also be eaten raw and the rich flesh retains its shape when cooked. AGM.

BRAMLEY'S SEEDLING Mid- to late season. PG3. A vigorous heritage and commercial variety; bright green skin and acidic flesh that cooks to a puree. AGM.

EDWARD VII Late season. PG6. Reliable trees have some disease resistance. Acidic apples cook to a puree. AGM

FLOWER OF KENT (syn. 'Isaac Newton' or 'Isaac Newton's Tree') Mid-season. PG3. Heritage variety, bred from the tree growing in the garden of the scientist Isaac Newton. Delicately flavoured flesh cooks to a puree.

GRENADIER Early season. PG3. Reliable trees; blossom has some resistance to frost. Fruits have green, slightly ribbed skin; tart flesh cooks to a puree. AGM.

HOWGATE WONDER Mid- to late season. PG4. Heritage variety; enormous fruits have orange/red skin and flesh that keeps its shape when cooked. AGM.

LANE'S PRINCE ALBERT Mid- to late season. PG3. Heritage variety with attractive, large fruits; good-flavoured flesh stays partially intact on cooking. AGM.

NORFOLK BEEFING Mid-season. PG3. Heritage variety. Firm flesh develops a rich, spiced flavour when slow-baked whole. Traditionally known as "biffins".

REVEREND W. WILKS Early season. PG2. Large pale-skinned fruits break down to a relatively sweet puree on cooking. A compact heritage variety.

'Blenheim Orange' is an eater and a cooker and thus saves space in the garden.

'Bramley's Seedling' is a classic cooking apple that's great for purees.

> **TOP TIP** IF THE BRANCHES OF YOUR APPLE TREE ARE PARTICULARLY WEIGHED DOWN WITH FRUIT IN AUTUMN, SUPPORT THEM BY TYING THE CENTRE OF EACH BRANCH TO THE TRUNK OR A STAKE. REMOVE THE STRINGS AFTER HARVESTING.

ALSO TRY

Crab apples (*Malus* species and cultivars) bear prolific harvests of cherry-sized apples in autumn and put on impressive displays of blossom in spring. The fruits are only eaten cooked, being especially used for a jelly to accompany savoury dishes, and are also a popular winter food for wildlife. Choose single-flowered varieties so that pollinators can also enjoy the blossom. Try *Malus* 'Butterball', which has bright yellow fruits; *M.* 'John Downie', which has an upright, conical habit and excellent autumn colour; or *M.* × *robusta* 'Red Sentinel', which forms a compact tree and holds its red apples until Christmas.

The vibrant bright red fruits of the crab apple tree are grown for cooking rather than eating raw.

MEDLAR *MESPILUS GERMANICA*

Fresh medlars are rarely found in the shops, so the only way to get hold of these strange-looking winter fruits is to grow your own and pick them fresh from the garden. They're rich in vitamin **C** and have a delicious flavour that's reminiscent of tart cooking apples and dates.

GROW FROM Potted or bare-root trees
HEIGHT AND SPREAD Up to 6 × 6m (20 × 20ft)
HARDINESS Hardy
IDEAL SOIL Moist but well-drained
POSITION/SITE REQUIREMENTS ☼ ☼
TIME TO MATURITY Two or three years

CALENDAR

	SPRING	SUMMER	AUTUMN	WINTER
PLANT			▓▓	▓
HARVEST			▓▓	

Through the year
Deciduous medlars make attractive garden trees, with large white spring blossom and good autumn colour.

PLANT

Medlars grow well in partial shade, but for the heaviest crops, plant them in full sun in a sheltered location. Exposure to cold winds can damage the blossom, reducing the harvest. Avoid planting medlars in frost pockets, and keep them at least 5m (16ft) apart from other trees. They're best grown as freestanding trees, not trained, and their spreading, sometimes squat habit allows some choice when it comes to the look of the tree. Either prune so that it develops a clear trunk before branching (see p.82), or allow the branches to form lower down and grow as a large bush, perhaps in a forest garden setting. Medlars are generally sold on the Quince A rootstock (see p.24).

Plant medlars in autumn and winter, before the ground freezes. Stake and water well after planting (see p.28), then spread a mulch of compost or other well-rotted organic matter around the base of the tree.

IN CONTAINERS Only trees that are marketed as "dwarf" or "patio" medlars are suitable for growing in containers; plant other varieties in large raised beds if you don't have open ground. Use a soil-based, peat-free compost and support with a stake.

Stake young trees to support them, checking the ties regularly and loosening as required.

The fruits are ready to harvest in late autumn, as the leaves turn.

Medlar 'Nottingham': the unusual, puckered base of medlars has earned them nicknames such as "dog's bottom".

Medlars have a spreading habit but need little pruning once mature.

GROW

Formative pruning helps to establish a good framework of branches (see *p.82*); mature medlars need only minimal pruning every winter (see *p.84*). Give the trees an annual mulch of compost or well-rotted organic matter mixed with a little wood ash. Water young trees as required and mature trees in dry spells. Give container-grown trees regular watering and a liquid feed fortnightly.

Lay medlars in a single layer on trays, not touching, to blet (see *right*).

HARVEST

Medlars are ready when the stalk parts easily from the tree. The longer they're left on the tree, the more flavour they'll develop. Harvest before the first frosts and ripen them inside. To draw out their deep flavours and reduce their tartness, the crunchy flesh of medlars needs to be ripened to the point of fermentation – a process known as "bletting" or "retting". Leaving the fruit on the tree until after a frost can speed up the ripening process (the cold temperatures start to break down the flesh) so that they need less or no bletting, but it can cause rot.

To blet medlars, dip the stalks in a strong salt solution to prevent rotting and lay on trays. Leave in a cool, dark, dry place for two to three weeks until soft. The fruit can then be eaten raw, or preserved (typically into a jelly). Medlars yield around 20kg (44lb) of fruit from a fully grown tree.

VARIETIES

Medlar trees are self-pollinating, so you only need to plant one. There's little to distinguish between the taste of the different varieties, but some have more upright or spreading forms. It's also possible to buy dwarf medlars on rootstocks designed to reduce vigour, but these aren't always named varieties.

DUTCH Forms a spreading tree with medium- to large-sized fruit.

IRANIAN Heavy harvests of small fruits that ripen well on the tree; bushy habit. Award of Garden Merit (AGM, see *p.23*).

LARGE RUSSIAN Very large fruits borne on pendulous, drooping branches.

NOTTINGHAM Upright, compact form that suits small spaces. Prolific harvests of medium-sized fruit; AGM.

ROYAL Trees have a semi-upright habit. Small to medium-sized fruits have an excellent, well-balanced flavour.

WESTERVELD Forms smaller, slightly weeping trees, and is the most common variety found as dwarf trees. Considered the best variety for making medlar jelly.

MULBERRY *MORUS NIGRA*

Fresh mulberries are impossible to buy in shops because the ripe fruits are so delicate and juicy they're difficult to transport and store. Mulberries can form large architectural trees, but by training them or planting a dwarf variety you can enjoy their gorgeous fruits from even a small garden.

GROW FROM Potted or bare-root plants
HEIGHT AND SPREAD Up to 10 × 5m (30 × 15ft)
HARDINESS Hardy
IDEAL SOIL Moist but well-drained
POSITION/SITE REQUIREMENTS ☼
TIME TO MATURITY Three years or more

CALENDAR

	SPRING	SUMMER	AUTUMN	WINTER
PLANT			▓▓	▓▓
HARVEST		▓ ▓		

Through the year
Large heart-shaped leaves (golden in autumn) and an architectural shape make mulberries an attractive garden tree all year round.

Freestanding trees in small spaces can be pruned in summer to restrict their size.

The dwarf 'Charlotte Russe' is the best choice for pot-grown mulberries.

PLANT

If you have the space, grow mulberries as freestanding trees in a sheltered, sunny spot, away from wherever frost settles in winter. Alternatively, train a tree as an espalier on a warm south- or southwest-facing wall, allowing for an ultimate width of 4.5m (15ft) and a height of 2.5m (8ft) (see p.86). Ideally, plant mulberries in autumn; you can plant in winter if the ground isn't frozen. Stake, water, and mulch the tree afterwards (see p.28).

IN CONTAINERS Wall-trained mulberries can be grown in a large raised bed at the foot of the wall, but for a pot, only dwarf varieties, such as 'Charlotte Russe', are suitable. Use a peat-free, soil-based compost and increase the pot size annually until the plant is fully grown.

GROW

Young trees need watering in dry spells, as they establish their roots in the first three to four years; but mature trees should need little watering. Water trees in containers and raised beds regularly and give them a liquid feed fortnightly through the spring and summer. Apply a mulch of compost or well-rotted organic matter mixed with a handful of wood ash in early spring every year.

After some formative pruning (see p.82), freestanding trees will need minimal, if any, pruning in winter, (see p.84). Prune espalier-trained trees in summer and winter (see p.89). The spreading branches of older freestanding trees can become brittle, and may need supporting with a forked stake. Check the ties on stakes regularly and loosen as appropriate.

Keep young mulberry trees well watered while the immature fruits are developing.

VARIETIES

All species of mulberry are self-pollinating (so only one tree is needed) and all flower quite late, escaping possible damage from the last frosts.

AGATE (*Morus alba*) Huge, black, and very sweet fruits. Trees are hardier than other species and have edible leaves.

CARMAN (*M. alba* × *M. rubra*) Sweet fruits, creamy-white in colour. Trees start producing fruit when relatively young.

CHARLOTTE RUSSE (syn. 'Mojo Fruit' and 'Waisei-kirishima-shikinari') A new dwarf variety that forms a compact (1.5m/4ft) bush rather than a tree (see *image, opposite*). Black fruit.

CHELSEA (syn. 'King James') This 17th-century heritage variety (see *p.23*) has very big, black fruits with a good sweet/acid balance.

ILLINOIS EVERBEARING Starts producing its black fruit when only two or three years old and crops for a long period.

WELLINGTON Red-black fruits are produced at a relatively young age. A variety from the USA, it ripens early in the season.

Most varieties ripen over a long period. Pick your mulberries regularly as the individual fruits ripen – daily, if possible.

HARVEST

Mulberries give frequent small harvests rather than gluts, so they're ideal for eating fresh. To preserve as jam, pick when slightly underripe. Fruits are ripe when fully coloured (white, red, or black, depending on variety) and plucked easily from the branch. They can be pulled off individually, but the juice stains fingers and fabrics – avoid this by laying an old sheet under the tree and shaking the branches so that the ripe fruits fall onto it. Yields are variable, but in a good year, expect 3–5kg (6½–11lb) from a young tree, and considerably more from a mature tree.

ALSO TRY

Amelanchier species are generally grown as ornamental trees, but many produce edible berries, also known as saskatoon berries, Juneberries, or serviceberries. Eat fresh or preserved. Most edible varieties are of *Amelanchier alnifolia*; try 'Smokey' or 'Thiessen' for reliable crops, 'Honeywood' for a more bushy plant, or 'Northline' for really juicy and flavourful berries. Juneberries grow as large, suckering shrubs or small trees, reaching 4m (13ft) high and 2m (6ft 6in) wide, and need little to no maintenance.

Juneberries are ready to harvest in midsummer, when the fruit is dark purple and slightly soft.

APRICOT *PRUNUS ARMENIACA*

A ripe apricot plucked from the tree has a depth of flavour unmatched by shop-bought fruits, which are picked underripe for ease of transport. With just a little effort to protect early blossom in spring, these low-maintenance trees will reward you with bowlfuls of glorious golden fruit in late summer.

GROW FROM Potted or bare-root trees
HEIGHT AND SPREAD Up to 4 × 2.5m (11 × 8ft)
HARDINESS Hardy
IDEAL SOIL Moist but well-drained
POSITION/SITE REQUIREMENTS ☼
TIME TO MATURITY Two years

CALENDAR

	SPRING	SUMMER	AUTUMN	WINTER
PLANT			▓ ▓ ▓	▓ ▓ ▓
HARVEST		▓ ▓		

Through the year
Fragrant white and pink apricot blossom is one of the highlights of the fruit garden in early spring.

PLANT

Most apricot trees are sold on the St Julien A rootstock, or the semi-dwarfing Torinel (see p.24). Plant in autumn in the sunniest, most sheltered spot you have, or against a south- or southwest-facing wall, or an interior greenhouse wall for fan-training (put the wires in first).

Make sure your apricot trees are spaced around 4m (13ft) apart; for fan-trained trees allow 1.8m (6ft) each way. Support freestanding trees with a stake. Water them well after planting and then mulch with compost or well-rotted organic matter.

IN CONTAINERS Dwarfing varieties reach 2m (6ft 6in) tall and can be grown in large pots filled with peat-free, soil-based compost. Keep in a greenhouse all year, or between late winter and late spring.

Protecting apricot blossom from frost is easier on a fan-trained tree than on freestanding trees.

Hand-pollinate apricot blossom on a sunny day when all the flowers are open.

GROW

Water young trees regularly and mature trees in dry spells, especially while flowering and fruiting. Water pot-grown apricots regularly; give a liquid feed fortnightly. Mulch with compost or well-rotted organic matter mixed with a handful of wood ash in late winter.

Prune apricots as for plums (see p.82 and pp.84–85) – pruning in summer helps protect them from silverleaf and bacterial canker (see pp.40–41).

Apricots' very early blossom and young fruit are vulnerable to frost. Protect with horticultural fleece at night, if possible (don't let it touch the flowers). Early flowering means there are often not enough insects around for successful pollination; those grown under cover will also need hand-pollination. To do this, use a soft paintbrush or feather to gently brush each flower and spread the pollen between the flowers. If there are lots of baby fruits, thin them to one per 5–8cm (2–3in) in early summer (see p.85).

Apricot **'Bergeron'** has pink blossom followed by reliably heavy crops of red-blushed fruits.

VARIETIES

Some varieties of apricot are juicier than others (and better for jam-making, for example) and the colour of the flesh and skin ranges from golden yellow to almost red. New varieties are being bred to be more tolerant of frost and cooler climates. All apricots are self-pollinating.

BERGERON Late season. French heritage (see p.23) and commercial variety used for jam; good frost resistance.

EARLY MOORPARK Mid- to late season. Juicy, yellow-skinned, and red-fleshed fruits with good flavour; UK heritage variety.

FLAVORCOT Early season. Heavy crops of large, juicy fruits on trees that are suited to cooler climates.

GARDEN APRIGOLD Early season. Compact trees (1.5m/5ft) in height and spread, golden fruits, and red-tipped foliage.

GOLDCOT Mid-season. Hardy, tolerating cold and wet well; golden fruits have relatively thick skins.

TOMCOT Early season. Very reliable; heavy crops of good-flavoured, large, orange/red fruits.

VIGAMA Late season. Flowers over several weeks, preventing fruit loss to frost damage. Aromatic orange/red fruits.

(see p.23)

ALSO TRY

Breeders have been crossing related *Prunus* species, with delicious results that have the aroma of apricots and the sticky texture of plums. Apriums are more apricot than plum – try *P.* 'Cot-N-Candy' (Aprium Series) – pluots are more plum than apricot, such as 'Purple Candy' and 'Flavor King', and plumcots, including 'Flavor Supreme, are a half-and-half fusion.

Hybrids fruits, such as 'Flavor King', are often larger than plums or apricots.

HARVEST

Apricots are ready in midsummer to early autumn, when fruits are aromatic, just soft, and part easily from the branch. They don't store well, so eat fresh or preserve soon after harvesting. A fully grown fan-trained tree will yield up to 14kg (31lb), a mature freestanding tree up to 55kg (121lb) in a good year.

Apricots start bearing fruit even when the trees are relatively young.

PEACH AND NECTARINE

PRUNUS PERSICA

A ripe, succulent peach or nectarine is a thing of beauty, which is why they often feature in art and literature. The introduction of dwarf cultivars suitable for pots has meant it's now even easier to grow your own and enjoy the sensuous pleasure of these gorgeous fruits ripened on the tree.

GROW FROM Potted or bare-root trees
HEIGHT AND SPREAD Up to 5 × 5m (8 × 15ft)
HARDINESS Hardy
IDEAL SOIL Moist but well-drained
POSITION/SITE REQUIREMENTS ☼
TIME TO MATURITY Two to three years

CALENDAR

	SPRING	SUMMER	AUTUMN	WINTER
PLANT	▨▨			▨▨▨
HARVEST		▨▨	▨	

Through the year Profuse pink blossom erupts from the bare branches of peach and nectarine trees during early spring.

PLANT

Both peach and nectarine trees are supplied on St Julien A rootstocks; grow them in the same way. They're best planted against a warm, sunny, south- or southwest-facing wall and trained as a fan (*see p.86*), spaced 3.5–5m (11–16ft) apart. Alternatively, you can fan-train a tree on the interior wall of a greenhouse or conservatory, planting the roots into the ground or a large pot. Buying a partially trained two- or three-year-old tree will give you a head start.

Plant in autumn, ideally, or in dormant season if the ground isn't frozen. Put in training wires before planting. Water well and mulch with compost or well-rotted organic matter after planting, and tie in the shoots to the wire framework.

IN CONTAINERS Dwarf varieties can't be trained and are designed for growing freestanding in pots inside or outdoors; they'll reach around 1.5 –1.8m (5–6ft). Plant into a large container (at least 50cm/20in in diameter) filled with peat-free, soil-based compost.

Training trees against a sunny wall or fence makes it easy to protect the trees and blossom from frost and rain.

Dwarf peach trees will thrive in a sunny courtyard.

Hand-pollinate each flower on a sunny afternoon, and repeat for the next two days.

GROW

Water young, mature, and potted trees through the growing season so the soil is consistently moist; irregular water supply can lead to deformities and splits in the fruit. Give a liquid feed to potted trees and those in the ground fortnightly in spring and summer. Mulch annually in late winter with compost or well-rotted organic matter and some wood ash.

Blossom on trees outside needs frost protection by covering with horticultural fleece. Keep rain off the plant in spring to help prevent peach leaf curl disease (see p.41). All trees, especially those under cover, benefit from hand-pollinating with a soft paintbrush. If there are lots of baby fruits, thin to 15cm (6in) apart when they're ping-pong-ball size (see p.84).

Fan-trained trees need formative pruning, and annual pruning and training thereafter in spring and summer (see pp.86–89); potted dwarf trees will need only dead or diseased wood removing every spring (see p.84).

Peach 'Bonanza' is a naturally dwarf variety that is ideal for containers.

VARIETIES

All peach and nectarine varieties are self-pollinating. Peaches have fuzzy skins, nectarines are smooth, but the flavour and texture of the fruits are broadly similar. The full botanical Latin name for nectarines is *Prunus persica* var. *nectarina*, peaches are simply *Prunus persica*.

PEACHES

AVALON PRIDE Late season. Vigorous trees suited to cooler climates; resistant to leaf curl. Red-fleshed fruit.

BONANZA Mid-season. Dwarf variety that bears reliable crops of good-sized, yellow-fleshed fruit.

PEREGRINE Mid-season. Reliable plants; richly flavoured, white-fleshed fruits with pale skins.

Beneath the red-flushed skin of 'Humboldt' is a rich yellow flesh.

ROCHESTER Mid-season. Large, golden, fibrous fruits with white flesh. Reliable outdoor crops.

NECTARINES

EARLY RIVERS Early season. Juicy fruit with a rich flavour and pale skin.

HUMBOLDT Mid-season. A heritage variety (see p.23) for growing under cover, it crops reliably heavily. Richly flavoured.

LORD NAPIER Mid-season. This heritage variety has reliable crops of fruits that have attractive, red-flushed skin.

NECTARELLA Early to late season. Reliable dwarf variety bearing sweet, well-flavoured fruits.

PINEAPPLE Late season. Heritage variety best grown under cover. Sweet fruit with pineapple undertones.

HARVEST

Peaches and nectarines are ripe and ready to pick between midsummer and early autumn, depending on the variety. The ripe fruit will be soft and aromatic, and part easily from the branch. Even with the best growing conditions, good harvests are not guaranteed each year. A mature fan-trained tree can yield as much as 5.5kg (12lb) or more of fruit in a good year; dwarf cultivars tend to yield somewhat less.

The fruits don't ripen all at the same time, so check them regularly.

PLUM AND GAGE

PRUNUS SPECIES

Plums are exceptional when homegrown: if picked only when ripe, they benefit from being able to develop their sugars and flavour to the full, unlike shop-bought fruit. When greengages and damsons are included in the mix, there's a wealth of flavours and textures to choose from that's rare in the shops.

GROW FROM Potted or bare-root trees
HEIGHT AND SPREAD Up to 3 × 3m (10 × 10ft)
HARDINESS Hardy
IDEAL SOIL Moist but well-drained
POSITION/SITE REQUIREMENTS ☼
TIME TO MATURITY Two to six years

CALENDAR

	SPRING	SUMMER	AUTUMN	WINTER
PLANT				
HARVEST				

Through the year Fragrant plum blossom is edible and can be used to decorate desserts – but the more you pick, the less fruit you'll get.

Damson, plum, and gage blossoms are vulnerable to late frosts.

PLANT

Ideally, plant bare-root and potted trees in autumn; you can plant in winter if the ground isn't frozen. Stake freestanding trees and water well after planting, then mulch with compost or well-rotted organic matter. The most widely available rootstocks are Pixy and St Julien A (see p.24) – use Pixy if you're going to train the tree as a cordon or fan, and put in the wires before planting. All plums, gages, and damsons prefer a warm, sunny, sheltered site; they'll grow on most soils, but for chalky soils choose a damson.

IN CONTAINERS Choose patio trees or those on dwarfing Pixy rootstock to grow as a branching tree or cordon (see p.18 and p.24). Plant and stake in a large container, at least 50cm (20in) across, filled with soil-based, peat-free compost.

GROW

Young trees and those in containers need regular watering; water mature trees during dry spells. Give container-grown trees a liquid feed fortnightly throughout the spring and summer. Mulch in late winter with compost or well-rotted organic matter mixed with a handful of wood ash.

Protect the blossom from frost if possible by covering with horticultural fleece. If there are a lot of fruitlets still on the tree after the June drop (see p.85), thin to avoid the branches breaking under the weight. Trees will need formative and then maintenance pruning during the summer (see pp.82–88); check the ties of trained and staked trees regularly and loosen or replace them if necessary.

Plums grow well as fan-trained trees against a warm south-facing wall.

Pick damsons when ripe but plums and gages slightly underripe, when using them for jam.

HARVEST

Pick when the fruit is fully coloured and slightly soft, and comes away easily from the tree; you'll need to return for multiple harvests. Take care you don't grasp a wasp as well as the fruit when picking – plums and gages are one of their favourite foods in late summer and early autumn.

Yields are variable, depending on the rootstock and variety, but can be 15–65kg (33–143lb) for a mature freestanding tree; 3.5kg (8lb) or more from a cordon; and 6.75kg (15lb) or more from a fan-trained tree.

The purple skins of plums are full of healthy anthocyanins.

ALSO TRY

Mirabelle plums (*Prunus cerasifera* and *P. insititia*) are also known as cherry plums for their small size. They produce heavy crops of deliciously sweet and juicy fruit earlier in the summer than plums and gages. Grow them in the same way as you would plums, choosing a self-pollinating variety if you've no other plum trees. 'Mirabelle de Nancy' is self-pollinating, with yellow fruit; 'Golden Sphere' is also yellow and 'Gypsy' is red – both of these are only partially self-pollinating.

Mirabelle plums make a superb jam and are also ideal for baking in tarts and pastries.

'Czar' plums are delicious eaten fresh from the tree, but they're also an excellent cooking plum (try them roasted).

The dark skins of 'Blue Tit' hide a golden yellow flesh best eaten raw.

'Early Laxton' is one of many heritage plums bearing the breeder's name.

VARIETIES

There is a huge variation in fruit flavour, texture, colour, and size across the plum family. Plums and gages are *Prunus domestica* varieties, damsons are *P. insititia* varieties. All dessert plums and gages can be cooked as well as eaten fresh, though they have varying levels of sweetness, while damsons are best cooked. If you plan to use your plums for jam-making, freezing, or other culinary purposes, consider choosing the easy-to-prepare "freestone" varieties, whose stones come away easily from the flesh.

Some varieties are self-pollinating ("SP" in the list, *see right*), others will need a partner tree in the same pollination group ("PG") to help them set fruit (see p.24).

PLUMS

AVALON PG2. Mid-season. Heavy yields of blush pink, freestone plums; good disease resistance.

BLUE TIT PG5; SP. Early to mid-season. Late-flowering, compact trees bear dark blue-purple fruits. Award of Garden Merit (AGM, see p.23).

CZAR PG3; SP. Mid-season. A heritage variety forming a compact tree; purple fruits are not too sweet. AGM.

DITTISHAM PLOUGHMAN PG3. Early to mid-season. Purple-red heritage variety that ripens within a short window; makes excellent jam; freestone.

EARLY LAXTON PG3. Mid-season. Compact trees bear sweet yellow-skinned fruit with a red blush.

EXCALIBUR PG5. Mid-season. Sweet and juicy fruits have a berry-like taste; yellow-red skin and yellow flesh.

HAUSZWETSCHE PG4. Late season. A damson-like plum that can either be eaten fresh when completely ripe or used for cooking.

MARJORIE'S SEEDLING PG5; SP. Late season. Trees have some canker resistance; deep purple skins and yellow flesh that isn't overly sweet. AGM.

OPAL PG3; SP. Early to mid-season. Reliable, heavy yields of round purple fruits on compact trees; freestone. AGM.

PERSHORE PG3; SP. Mid-season. Heritage variety with yellow skin and flesh; good for bottling and cooking.

VICTORIA PG3; SP. Mid- to late season. Popular heritage variety; tastier when cooked, retains its rosy colour on cooking, unlike other varieties. AGM.

WARWICKSHIRE DROOPER PG2; SP. Mid- to late season. Heritage variety with a weeping, spreading habit and yellow fruit.

GAGES

CAMBRIDGE GAGE PG4. Late season. Reliable heritage variety with green, sweet, richly flavoured fruits. AGM.

DENNISTON'S SUPERB PG2; SP. Mid-season. Reliable crops of green-skinned fruit with transparent flesh.

EARLY TRANSPARENT GAGE PG4; SP. Sweet, melting flesh; translucent green skins have a red blush. Heritage variety good for cooking.

IMPERIAL GAGE PG2; SP. Mid-season. Yellow-green fruits with juicy flesh; a reliable heritage variety. AGM.

OULLINS GAGE PG4; SP. Mid-season. Reliable trees have some canker resistance; rosy blush to the green fruits. AGM.

'Cambridge Gage' has small fruits with a honeyed flavour.

DAMSONS

BLUE VIOLET DAMSON PG3; SP. Mid-season. Heritage variety; sweet, plum-like fruits fall when ripe.

FARLEIGH DAMSON PG4; SP. Late season. Heavy cropping but compact heritage variety; excellent for jam-making. AGM.

MERRYWEATHER DAMSON PG3; SP. Late season. Heavy crops of large, mildly acidic fruits; vigorous trees with good disease resistance.

PRUNE DAMSON (syn. 'Shropshire Prune') PG3; SP. Late season. Heritage variety excellent for cooking; reliable crops of dark, firm fruits. AGM.

'Prune Damson' is intensely flavoured and perfect for making damson gin.

When fully ripe, 'Merryweather' damsons can be eaten fresh.

CHERRY *PRUNUS* SPECIES

Whether you prefer a luscious, sweet dessert cherry or the sour tang of acid cherries, homegrown cherry harvests are a highlight of summer. Acid cherries are found in the shops only in tins or jars because the fresh fruits don't store well, so growing your own is the only way to get hold of them.

GROW FROM Potted or bare-root trees
HEIGHT AND SPREAD Up to 8 × 8m (26 × 26ft)
HARDINESS Hardy
IDEAL SOIL Moist but well-drained
POSITION/SITE REQUIREMENTS ☼ ☼
TIME TO MATURITY Three years

CALENDAR

	SPRING	SUMMER	AUTUMN	WINTER
PLANT	▨		▨	▨
HARVEST		▨		

Through the year
Cherry trees put on a show of brilliant autumn colour as well as their profuse spring blossom.

PLANT

Most cherry varieties are supplied on restricting rootstocks: allow 8m (26ft) between the trees on Colt rootstocks and half that distance for those on Gisela 5 rootstocks (see p.24). Train them as fans (see p.86) or grow as freestanding trees. Dessert and Duke cherries need full sun; acid cherries can be grown in partial shade, including north- or east-facing walls.

Autumn is the best time to plant – or winter, if the ground isn't frozen. Put in wires for trained forms first. Stake freestanding trees and water well after planting, then mulch with compost or well-rotted organic matter.

IN CONTAINERS Choose patio trees or those on dwarfing Gisela 5 or Tabel rootstocks (see p.18 and p.24). Plant into a large container at least 50cm (20in) across, filled with soil-based, peat-free compost, and stake.

Cherry-tree fans can be squeezed into small gardens and are more easily protected from frost and birds.

Cherry blossom is widely celebrated, especially in Japan, but the plant is even more glorious when followed by fruit.

GROW

Water young trees and those in pots regularly. Give pot-grown trees a liquid feed fortnightly through spring and summer. Water mature trees in very dry spells and mulch in late winter with compost or well-rotted organic matter mixed with a handful of wood ash.

Protect the blossom from frost if possible by covering with horticultural fleece. Trees will need formative and then maintenance pruning in summer (see pp.82–88); check the ties of trained and staked trees regularly and loosen or replace them if necessary. Deter birds from the ripening fruit, if desired (see pp.44–45).

'Stella' is a reliable dessert cherry popular with commercial growers.

HARVEST

Pick cherries when they're deeply coloured, soft, and sweet. They don't store well once picked, lasting only one to three days. To preserve, freeze them (see p.46), or cook in desserts, ice-cream, or jam. Yields for a mature freestanding cherry tree can be 13.5kg (30lb) or more; expect at least 5.5kg (12lb) from a mature fan.

> **TOP TIP** FOR MAXIMUM HEALTH BENEFIT, EAT CHERRIES AS SOON AS POSSIBLE AFTER PICKING, AS THEIR NUTRIENT CONTENT STARTS TO DEGRADE WHEN THEY'RE OFF THE TREE.

'Morello' acid cherries are ideal for cooking when they turn red, but after further ripening on the tree to a deep crimson they can be eaten fresh.

VARIETIES

Cherry varieties are divided into sweet dessert types (*Prunus avium*); acid (sour) cherries (*P. cerasus*); and Duke varieties (*P. avium*), which are a cross between sweet and acid cherries. Acid cherries contain similar levels of sugar as the sweet varieties but, balanced by sourness, they have more flavour and are good for jam and baking.

Some varieties are self-pollinating (abbreviated as "SP" in the list below); others need a partner tree in the same pollination group ("PG" below, see also p.24) to help them set fruit.

CELESTE (syn. 'Sumpaca') Mid-season. SP, PG4. Naturally dwarf trees produce very sweet, dark red cherries.

EARLY RIVERS Early season. PG1. A vigorous heritage variety (see p.23) with large, delicious fruits.

KORDIA Late summer. PG4. Dark purple-red fruits with good acid/sweet balance. Award of Garden Merit (AGM, see p.23).

MAY DUKE Mid-season. Partially SP, PG3. A heritage Duke cherry with superb flavour.

MONTMORENCY Early season. SP, PG5. This acid cherry is popular in the USA and makes the best cherry pie.

MORELLO Late season. SP, PG4. The archetypal acid cherry; heritage variety with excellent flavour. AGM.

NABELLA Late season. SP, PG5. Heavy yields of dark acid cherries on relatively compact trees.

STELLA Mid-season. SP, PG4. Sweet, deep red dessert cherry that is a good pollinator for other varieties. AGM.

VEGA Mid-season. PG3. Large, white dessert cherries evade the attention of the birds.

WATERLOO Mid-season. SP, PG2. Delicious, heart-shaped, heritage dessert variety with a small stone.

PEAR *PYRUS COMMUNIS*

As with many fruit trees, pears can become quite large if grown freestanding in the ground, but you can also grow these delicious fruits in all sorts of spaces by training them onto walls, fences, or posts and wires, or by growing in pots. Their early blossom is also very beneficial for insects.

GROW FROM Potted or bare-root trees
HEIGHT AND SPREAD Up to 4 × 8m (13 × 26ft)
HARDINESS Hardy
IDEAL SOIL Moist but well-drained
POSITION/SITE REQUIREMENTS ☼
TIME TO MATURITY Three years or more

CALENDAR

	SPRING	SUMMER	AUTUMN	WINTER
PLANT	▨▨			▨▨
HARVEST		▨	▨▨	

Through the year Trained pear trees add aesthetic structure to the garden and are highly decorative at all times of year.

Trained pears, such as these three double cordon trees, are ideal for edging paths and borders.

PLANT

Pears are available primarily on Quince A or Quince C rootstocks (see p.24). They flower early in spring, and benefit from being planted and trained as cordons or espaliers (see pp.86–87) on south- or southwest-facing walls to protect the blossom from frosts. Put training wires and posts in first and plant in autumn (ideally) or winter, but not into frozen ground. Space freestanding trees and espaliers 4m (13ft) apart, and cordons 75cm (30in) apart. After planting, tie into the wires, water well, and then apply a mulch of compost or well-rotted organic matter.

IN CONTAINERS Plant dwarf pears or those on a Quince C rootstock into large pots (at least 50cm/20in in diameter) filled with soil-based, peat-free compost. Grow as freestanding cordons or trees (support with a stake), or put the pot at the base of a wall and train as above.

Pots with handles can be easily moved inside to protect the blossom from frost.

An espaliered pear tree gives large harvests but takes up little space.

GROW

Young trees and those in pots need watering regularly; water mature trees in very dry spells. Mulch trees in late winter with compost or other well-rotted organic matter mixed with some wood ash; give pot-grown trees a liquid feed fortnightly in spring and summer. Trees need formative and maintenance pruning and training onto wires, if required (see pp.82–88); prune when trees are dormant in winter. Protect blossom from frost with horticultural fleece and thin fruits in early summer to two per cluster with about 10cm (4in) between clusters. Look out for codling moth and canker (see p.39 and p.40).

The oval-shaped fruits of 'Williams' Bon Chrétien' are delicious eaten fresh but also preserve and poach especially well.

VARIETIES

Pears are grouped into dessert and culinary varieties. All need planting with another compatible variety for good pollination; the pollination groups ("PG") are shown in the list below (see also p.24). Alternatively, for several compatible pear varieties on one tree, purchase a "family" tree (see p.48).

BLACK WORCESTER Mid-season. PG3. Culinary heritage variety (see p.23) from the UK; delicious purple-flushed, rough-skinned fruits.

CATILLAC Mid-season. PG4. French heritage culinary variety; vigorous trees produce large fruits.

CONCORDE Mid-season. PG3. Compact trees produce reliable crops of juicy dessert pears. Award of Garden Merit (AGM, see p.23).

CONFERENCE Mid-season. PG3. Reliably heavy crops of dessert pears even in cooler areas. AGM.

DOYENNE DU COMICE Late season. PG4. French heritage eating variety with excellent flavour and red-flushed skin. AGM.

OBELISK Mid-season. PG3. A dwarf variety with a columnar habit ideal for growing in pots.

WILLIAMS' BON CHRETIEN Early season. PG3. Heritage dessert variety; juicy and smooth-skinned pears, compact trees.

HARVEST

Summer varieties can be eaten straight from the tree, but autumn pears will need picking unripe. Pick by holding the fruit in your palm, with your index finger on the stalk end. When you tilt the pear upwards it will detach easily from the branch if it's ready. Store on trays in a cool, dry place, ensuring they're not touching each other. Bring into a warm room to ripen them; when ripe, they'll be soft around the stalk end.

A single mature cordon will yield up to 3.5kg (8lb) of fruit, an espalier 9kg (20lb) or more, and a freestanding tree at least 18kg (40lb).

ALSO TRY

Grow aromatic Asian pears (*Pyrus pyrifolia*) in the same way as pears. You'll need two trees (or one Asian pear and one standard pear) with compatible pollination groups as they're not self-pollinating. Try varieties 'Nashi Kumoi' (PG 2, 3, or 4) or 'Sensation' (PG3). Harvest when fully ripe as they won't ripen further indoors.

Asian pears have crisp, firm flesh suited to using in salads or with cheese.

ELDER *SAMBUCUS NIGRA*

Elder isn't a traditional fruit crop, being more often foraged, but its tendency to grow near roadsides means the flowers and fruit can be contaminated by pollutants, and those in good locations can be stripped by other foragers before you arrive. Guarantee good-quality fruit by growing your own.

GROW FROM Potted plants
HEIGHT AND SPREAD Up to 8 × 4m (26 ×13ft)
HARDINESS Hardy
IDEAL SOIL Moist but well-drained
POSITION/SITE REQUIREMENTS ☼ ☼
TIME TO MATURITY Three years

CALENDAR

	SPRING	SUMMER	AUTUMN	WINTER
PLANT				
HARVEST				

Through the year
Elder gives a double harvest of edible flowers in spring and fruit in late summer and autumn.

Elderberries will grow in most situations, but will fruit best in full sun.

PLANT

As evidenced by their ubiquity in scrubland and woodland, elders aren't fussy about their growing conditions and are ideal to plant in corners where nothing else will survive. If planting more than one, space them 3m (10ft) apart.

Plant in autumn or winter, provided the ground is not frozen, and water well after planting. Spread a mulch of compost or well-rotted organic matter around the base.

IN CONTAINERS Elders are not suitable for growing in containers.

GROW

Water young plants regularly, and mature plants during very dry spells. Mulch in late winter with compost or well-rotted organic matter mixed with a little wood ash.

Any pruning will reduce the harvest that year, but you can cut the stems back to keep elder from getting too big. In early spring, either prune as for blackcurrants (see p.115) or cut the whole plant back to ground level or a short stump. It will regrow by around 1.8m (6ft) in a single year.

Hard pruning to a short stump keeps plants to size, but yit means you'll lose that year's harvest.

Purple-leaved and pink-flowered elders such as 'Black Lace' are an ornamental alternative to the species.

> **TOP TIP** ELDERBERRIES ARE RICH IN VITAMINS A AND C, FLAVONOIDS, AND ANTIOXIDANTS. MODERN RESEARCH NOW SUPPORTS THEIR AGE-OLD USE AS AN IMMUNE-BOOSTING TONIC IN AUTUMN.

VARIETIES

It's best to grow the species for fruit (*Sambucus nigra*), but a number of more ornamental varieties, the finest of which are listed below, can also be used. All species are self-pollinating, although having more than one plant will result in heavier yields.

AUREOMARGINATA Leaves of this variety are variegated, each one with an irregular creamy-white edge.

BLACK LACE (syn. 'Eva'; *S. nigra* f. *porphyrophylla*) Finely dissected purple foliage and pale pink flowers. Award of Garden Merit (AGM, *see p.23*).

LACINIATA (*S. nigra* f. *laciniata*) Fern-like green leaves; this form is slightly smaller than the species. AGM.

HARVEST

In spring, elder bushes are covered with frothy, fragrant, edible flowers that are used to make drinks and desserts; but the more you pick, the fewer berries you'll get. Elderberries are ready to pick from late summer, when coloured

Elderberries can be made into jam, syrup, and cordial, or cooked with other fruits in pies and crumbles.

a deep purple and the sprig is hanging pendulously on the bush; those in shade will ripen later than bushes in full sun. Pick the whole sprig (the stalk will snap easily from the stem); once home, run the sprig through the tines of a fork to remove the berries. Yields depend on the size of the bush and how many flowers were picked, but a mature bush can easily produce several kilos.

Berries must be cooked; do not eat them raw. Leaves are toxic if eaten.

ALSO TRY

Honeyberries are fruiting honeysuckles producing elongated, almost cuboid, berries. These climbers will reach around 1.2m (4ft) or more in height and spread, and need only minimal pruning. Note that not all honeysuckle berries are edible. Plant and pick only named edible species, which are: *Lonicera caerulea* var. *edulis* and *L. caerulea* var. *kamtschatica* – the latter is sometimes sold as *L. kamtschatica*, of which there are a number of varieties available.

Honeyberries are antioxidant like elder, but have a lighter, more floral taste.

PRUNING YOUNG FRUIT TREES

Most fruit trees naturally produce irregular clusters of crossing branches, which bear less fruit than trees with an open, balanced branch structure. Formative pruning helps shape a young tree into one that will be strong, productive, and easy to maintain. Depending on the age of the tree you buy, some of the pruning may have been done by the nursery growers, but it's a straightforward and quick job to tackle.

Remove unwanted branches back to the main stem.

CREATING A GOBLET-SHAPED TREE

Freestanding fruit trees are most often grown into a goblet-like form. These have an open-centred crown of branches on a clear trunk. Variants of this goblet form are known by different names depending on the height of the trunk before it starts to branch: these technical designations are: bush, 75cm (30in); half-standard, 1.35m (4ft 5in); and standard, 2m (6ft 6in). Most fruit trees, including apples, pears, cherries, plums, apricots, peaches, persimmons, and quinces, can be pruned to this shape.

If you've bought a maiden whip tree, prune it after planting to the desired height by cutting just above a bud (*below left*); make sure that there are a few plump buds just beneath it. The point at which you cut will be the height of the branching – if it's too short, grow it on for a year before pruning to height. Once the tree has branches – or after planting if you've bought a feathered maiden (*below centre*) – cut the branches back to one-third of their length and the central stem back to just above the topmost branch. The following year (*below right*), select three or four of the strongest and best-placed branches to form your main goblet framework, shortening these to one-third of their length; remove any misplaced (e.g. crossing) branches back to the stem and shorten others to four buds from the stem.

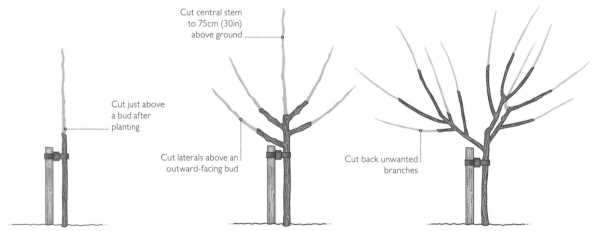

Cut central stem to 75cm (30in) above ground

Cut just above a bud after planting

Cut laterals above an outward-facing bud

Cut back unwanted branches

Prune a maiden whip to the desired height – typically around 75cm (30in) from the ground – after planting.

A feathered maiden should have its central stem removed and the side branches shortened by two-thirds.

In the following year, select and shorten the branches for the main framework and remove or shorten the remainder.

PYRAMID-SHAPED TREES

Plum, apple, pear, peach, and apricot trees can also be pruned into an attractive pyramid (cone) shape, but this form requires the support of a permanent stake. To create a pyramid from a feathered maiden, tie in the tree to a 2m (6ft 6in) stake after planting. Shorten the central stem to around 15cm (6in) above the topmost branch and shorten the branches by half, removing any below 45cm (18in) from the ground.

In the following year, prune the new growth back to about 20cm (8in) for the main branch, and any new growth off the main branch to 15cm (6in): always try to prune to a downward-facing bud. Shorten the central stem by two-thirds of the new growth. Repeat this for the following years until the tree is the desired height, adjusting the ties as necessary.

NEED TO KNOW

Fruit trees are sold as one-, two-, and three-year-olds. A one- or two-year-old is ideal if you want a freestanding tree, but if you prefer a trained tree, a three-year old with initial training is preferred, though not essential. Older trees are more expensive but need less formative pruning and bear fruit sooner.

- A maiden whip is a one-year-old tree that has a central stem (trunk) but no side branches.
- A feathered maiden is a two-year-old tree with side branches – look for one that has several strong-looking (not spindly), evenly spaced branches at a V-shaped angle from the central stem.
- Part-trained trees will be three years old and have had some of their lower branches tied down onto a frame of canes, ready to transfer to your garden as a young espalier or fan (see pp.86–87).

A fruit tree pruned as a pyramid is an attractive structural alternative to the classic goblet shape.

PRUNING MATURE FRUIT TREES

As fruit trees mature, they need a little attention each year to stay healthy and productive (pp.82–83 explain the first three years of pruning; this section picks up from there). Mature stone fruit trees, such as plum and cherry, should have their main pruning in summer, whereas apple and pear trees are pruned in winter, when the tree is dormant. Refer to the directory (*pp.50–81*) for species-specific guidelines.

Medlar trees need little pruning beyond the removal of the four Ds.

Branches that are crossing will rub and damage each other, so remove the worst-placed of the two.

THE FOUR Ds

The "four Ds" are a great shorthand to help you remember the main targets for pruning. They refer to dead, diseased, dying, and duplicate branches. The first three are self-explanatory, while "duplicate" refers to cases where two branches are growing in the same direction close to one another. The upper will shade out the lower, so the weaker should be pruned out. Always remove the four Ds before undertaking any other pruning. For some species, such as medlar, quince, and mulberry, that will be all the pruning that is needed.

PRUNING APPLE AND PEAR TREES

After dealing with the four Ds (see *left*), the next step in creating a goblet-shaped tree (see *p.82*) is to shorten or remove branches that are crossing or growing up into the centre of the crown: the aim is to keep this area clear to allow air and light to reach all branches. Fruit on crowded trees won't ripen well and higher humidity in dense branching fosters disease. Shorten the main and secondary branches as necessary, then remove older branches to keep the framework balanced. Shorten sideshoots off the main and secondary branches; these will form "spurs", short sections of branch that bear fruit. The more vigorous the tree, the lighter the winter pruning you should do, as it stimulates new growth. Some apple and pear varieties bear fruit on the tips of wood produced the previous year. Notable examples of such "tip-bearers" are 'Bramley's Seedling' and 'Blenheim Orange'. Prune them in winter following the guidelines shown right.

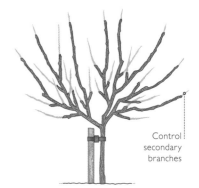

Control secondary branches

In the fourth winter of an apple or pear, extend the framework of secondary branches by removing the top third of main branches and taking the secondary branches back to four buds long (or two if they're spindly). Remove any crossing branches or those growing into the centre.

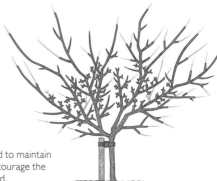

A mature tree is pruned to maintain its overall shape and to encourage the production of fruiting wood.

PRUNING OTHER TREES

Prune cherry, plum, apricot, peach, persimmon, citrus, and freestanding fig trees in summer. First, remove any of the Ds as required, then prune out only the least productive and worst placed branches (such as those growing into the centre of the tree) to maintain a balanced framework. Long secondary branches can be shortened to encourage the growth of more branches or, if the tree appears congested, removed altogether. Lightly prune the ends of the main branches, without removing too many buds.

Thinning out the weakest fruits allows the rest to develop fully.

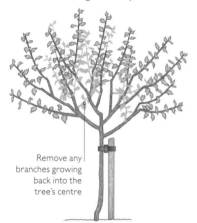

Remove any branches growing back into the tree's centre

Mature stone fruit trees don't respond well to heavy pruning, so remove only what's necessary.

TOP TIP SOMETIMES A TREE WILL PRODUCE MORE BABY FRUITS THAN IT CAN SUPPORT. THE TREE WILL THIN THESE OUT ITSELF, A PROCESS KNOWN AS "JUNE DROP", BUT YOU CAN THIN THEM FURTHER AFTERWARDS. REMOVE ANY MISSHAPEN FRUITS OR THOSE THAT HAVE NO SPACE TO DEVELOP; THIS ALLOWS THE TREE TO PUT ITS ENERGY INTO GROWING THE REMAINING FRUITS TO THEIR FULL SIZE AND BEST FLAVOUR.

PRUNING PYRAMID-SHAPED TREES

The main aim of pruning an established pyramid tree (after removing the four Ds) is to maintain its shape, which is not only attractive, but also allows for easy picking of fruit. Mid-spring is the time that you should remove or shorten any branches that are starting to spoil the overall look of the tree.

A pyramid shape ensures that more leaves and fruit gain exposure to the sun.

Cut back vigorous vertical shoots in summer

NEED TO KNOW
- Never take out more than a quarter of a tree's branches in a single year.
- The harder the pruning, the stronger the tree and the stronger the regrowth. Overly hard pruning stimulates the growth of upright branches that will crowd the crown.
- If a tree starts biennial bearing (a heavy crop one year, little the next), ensure it has enough water and nutrients. Thin the fruits heavily in the cropping year to rebalance.

TRAINING FRUIT TREES

Training fruit trees is a simple and highly effective way to grow fruit in a small space. What's more, the dramatic, architectural structure of trained trees can enhance the visual appeal of your garden. Cordons, espaliers, and fans can be trained against vertical surfaces or used to create a bountiful living screen, while stepovers (horizontal cordons) are a pretty way to edge a bed that also provides a harvest.

Training fruit against a flat surface is particularly appropriate for early-blossoming fruits such as peaches that benefit from enhanced shelter.

Fix eye bolts into a wall, fence, or post and connect the wire via expanding bolts that allow the wire to be tensioned.

TREE SUPPORT

To train a tree into a desired shape, the pliable young growth needs to be tied, and regularly retied, to horizontal wires. Even after the wood has hardened and the branches "set", they still need support from the wires.

Before planting, fix galvanized wires into place on a fence, wall, or between sturdy freestanding posts. The wires should be 4–10cm (1½–4in) from the surface. Space them 45cm (18in) apart vertically for fan-trained trees and 60cm (24in) apart for cordons and espaliers, to as high as you want the tree to be. Stepovers need only one wire about 45cm (18in) from the ground between posts 3–4m (10–13ft) apart.

CORDONS AND STEPOVERS

A cordon may be a single stem, grown upright with the support of a cane in a pot, or trained against a wall at an angle (an "oblique cordon"). Alternatively, it can be trained like a two-pronged fork ("multiple cordon"). Cordons are suitable for apples, pears, and plums (oblique cordons only).

If you're aiming for an oblique cordon, tie the tree to a cane after planting and cut back any sideshoots to three buds long (see pp.88–90 for subsequent pruning). Start a multiple cordon as you would an espalier (see right), pruning at the first wire, then tie in two branches to grow vertically.

Stepovers are the smallest of trained fruit trees – essentially one-level espaliers grown on miniature rootstock. Plant apple or pear stepovers 1.5–2m (5–6ft 6in) apart, supporting each one support the vertically with a cane tied to the wire. The following spring, carefully bend the stem over to the wire and tie it down in several places, then prune in summer (see pp.88–90).

Single cordons against a surface are best planted at a 45° angle to promote better fruiting; a bamboo cane tied into the wire at the same angle gives ample support.

Use natural twine to tie cordons; eventually, it will break down safely.

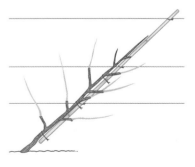

ESPALIER

Espalier training, in which branches are trained symmetrically in horizontal tiers along wires, is suitable for apples, pears, and mulberries. If you have bought a partially trained tree, tie the trained branches to your own wires after planting and remove the bought framework. If you begin with a maiden whip, tie it in to the lowest wire and prune to a bud just above the tie. The branches are lowered to the horizontal in two stages. In the summer, tie in two sideshoots at 45° angles. The following autumn, bend them down and tie to the wires below. Repeat this process in the following years (see also pp.88–90).

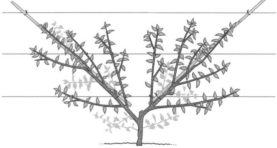

Apple espaliers generally need little pruning once established: you should cut back any sideshoots from the horizontal leaders to three leaves.

TOP TIP BUYING AN OLDER, PARTIALLY TRAINED TREE WILL GIVE YOU A HEAD START ON TRAINING ESPALIERS AND FANS; OR USE A FEATHERED MAIDEN OR MAIDEN WHIP FOR ALL TRAINED FORMS.

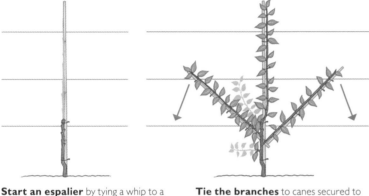

Start an espalier by tying a whip to a cane and pruning to the first wire.

Tie the branches to canes secured to the wires and lower to 90° in two stages.

FAN

The fan form is particularly suitable for cherries, plums, apricots, peaches, nectarines, and fig trees. Fan trees are best begun with a feathered maiden, because that provides you with the first stems to tie in. Remove the central stem and all but two equal-sized sideshoots around the level of the first wire; use canes tied to the wires to support the two selected shoots. Tie these shoots in at 45° angles and prune them back by two-thirds. In the summer, tie in the new growth of the two original shoots and select about six well-spaced sideshoots on each one. Tie the new shoots onto the horizontal wires to create the fan shape, remove any shoots originating from beneath the two principal arms of the fan, and cut back unwanted sideshoots to a single leaf. Early in the following spring, cut each shoot back by a third and remove any central growth.

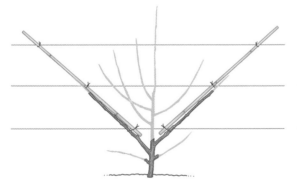

Prune a feathered maiden in the first spring after planting, removing the central stem and all but two sideshoots.

The following summer, tie in the lengthened sideshoots and new shoots that you want to keep to develop the fan shape.

PRUNING TRAINED FRUIT TREES

Pruning a trained tree take a little more time and attention than is needed for a freestanding tree. Devoting an hour or so to a tree once or twice a year will maintain its sharp structural form and keep it healthy and productive. Trained fruit trees are pruned in summer and sometimes in winter or early spring: summer pruning restricts growth, while pruning when the branches are bare promotes it.

Apple and pear are the most popular espalier trees because their fruiting spurs can produce good yields for many years.

PRUNING CORDONS AND STEPOVERS

The main – and usually the only – pruning required for cordons and stepovers is carried out in summer. Shoots from the main stem typically have a cluster of buds at their base, then a section where the leaves are more spaced out. Cut back each of these shoots to three leaves from the stem, not including the basal cluster. However, any shoots that are growing directly towards the wall or fence behind the cordon should be cut off at their base. Shoots that grow off the sideshoots should be shortened to one leaf from their base. Leave the central stem to keep growing until it has reached the height (or length, for stepovers) that you desire, and thereafter cut it back to that length each summer. Winter pruning is not usually necessary for cordons and stepovers, but on older trees you can thin out the most congested and least productive spurs (fruiting sideshoots) and any of the Ds (see p.84). Treat each stem of a multiple cordon as a single cordon, and cut back any growth from the centre to its base (or allow it to grow up as a central, third prong of the fork shape).

Training fruit trees can break up expanses of wall, adding visual interest to your outdoor spaces.

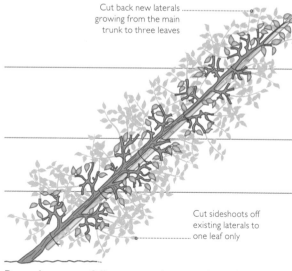

Cut back new laterals growing from the main trunk to three leaves

Cut sideshoots off existing laterals to one leaf only

Removing excess foliage on cordon trees keeps the tree to size and helps the sun reach the fruit to ripen it.

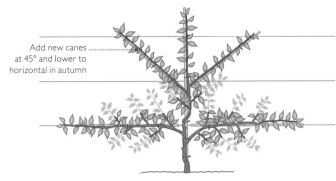

Add new canes at 45° and lower to horizontal in autumn

In an espalier's second summer, prune sideshoots from the lower arms back to three leaves above the basal cluster of buds.

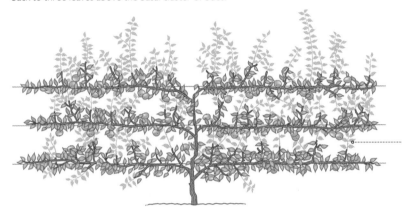

PRUNING ESPALIERS

Continue tying in the central stem of the espalier and adding a new tier of branches every year (see pp.86–87) until all the wires are occupied with horizontal branches. Each summer prune the sideshoots on each branch back to three leaves long (not including the basal cluster of buds), and prune any shoots off the sideshoots to one leaf long. In winter, remove the Ds (see p.84) and any ingrowing shoots, and shorten the central stem to a bud just above the next wire. When you've finished training all the tiers, shorten any shoots from the centre to three leaves in summer.

Cut back laterals growing from the main arms to three leaves

On an established espalier, treat each branch in the same way that you would treat an established cordon (see opposite).

PRUNING FANS

Mature fan-trained trees need to be pruned three times during the year. In early spring, once the tree is in growth, cut back the central branches by a quarter to encourage further branching: do this every year until the fan has filled its allotted space. Thereafter, use spring pruning to maintain the health and structure of the fan: this means removing the Ds (see pp.84) and any badly placed shoots, including those growing towards the wall. Thin out older fruiting spurs as necessary.

In early summer, tie in any shoots you want to retain for the framework of the fan and cut back those you don't to six leaves from the base. After the harvest in the autumn, cut these shoots back again, this time to just three leaves from the base.

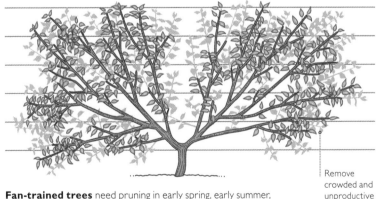

Remove crowded and unproductive growth

Fan-trained trees need pruning in early spring, early summer, and then in early autumn to keep them to shape.

NEED TO KNOW
- In a trained tree, horizontal or angled branches will produce more flowers and fruit (rather than shoots and leaves) than vertical ones.

- Fruit on some trained trees can need thinning after the June drop (see pp.85). Trees that might need such thinning are apple, pear, plum, apricot, peach, and nectarine.

A blackcurrant bush is a terrific low-maintenance plant that will produce good harvests of fruit as well as flavoursome leaves that can be infused into sorbets and herbal drinks.

FRUIT BUSHES AND PERENNIALS

Growing most berries, currants, and perennials is incredibly straightforward. Some, such as blackcurrants, can take less than an hour's maintenance each year but will reward you with copious quantities of tasty fruits every summer.

SWEET TASTE

To develop their flavour to full potential, berries and currants should be picked only when fully ripe. Growing your own fruit means you can wait until they're at their absolute peak. Ideally, pick them in the afternoon once they've been warmed by the sun. The berry family is far larger than you'd imagine from browsing the produce on supermarmarket shelves. Growing your own allows you access to fruit such as lingonberries, Chilean guavas, and humble gooseberries that are hard to find in the shops. You can also choose varieties of popular fruits that are more flavourful. For example, blueberries bred for supermarkets now have virtually white flesh, but you can grow a variety that has purply-blue flesh and therefore contains more healthy antioxidants.

GREAT SAVINGS

Berries and currants – known as "soft fruits" – and even the hardy rhubarb and tropical pineapple are all relatively labour-intensive for farmers to harvest, and are therefore some of the more expensive fruits to buy in the shops. Inevitably, because they're delicate, berries are sold in small quantities and plastic punnets. However, a single mature blueberry bush can yield up to 5kg (11lb) of fruit, with no single-use plastic required, and no food miles.

IDEAL CONDITIONS

Excluding pineapples (which, in temperate climates, need to be grown under the cover of polytunnels or greenhouses) and the few that need acidic soil, fruit bushes and perennials are a tolerant group of plants and some of the easiest and lowest-maintenance of all fruit plants to grow.

They'll fruit best in a sunny, kitchen-garden setting. Some of these plants – such as gooseberries and redcurrants – can also be trained against walls or fences, or used as edible edging plants, such as Chilean guava. All fruit bushes and perennials can be grown in pots. Most need only minimal pruning, and some, such as rhubarb, don't need any at all.

PINEAPPLE *ANANAS COMOSUS*

Pineapples are tropical plants that will grow quite happily indoors in temperate climates. Reduce food waste and food miles by growing your own from the top of a shop-bought fruit. The leafy growth of this bromeliad also makes an eyecatching, attractive houseplant.

GROW FROM Pineapple tops or potted plants
HEIGHT AND SPREAD Up to 1 × 1m (3 × 3ft)
HARDINESS Tender
IDEAL SOIL Well-drained
POSITION/SITE REQUIREMENTS ☼
TIME TO MATURITY Three years or more

CALENDAR

	SPRING	SUMMER	AUTUMN	WINTER
PLANT				
HARVEST				

Through the year
Evergreen pineapples can crop at any time, but are most likely to flower in the spring or summer.

Repot your pineapple plant when it outgrows its container, putting it into a slightly larger one each time.

PLANT

To root your own plant, cut off the fruit's leafy top, then strip off some of the lowest leaves to leave a bare stub of bumpy leaf scars at the base, around 5cm (2in) long (trim off any flesh below the scars). Leave it to dry for two days. Suspend the top over a narrow-necked jar of water so that only the base is submerged. Keep on a sunny windowsill until it starts to produce roots, which could take several weeks.

IN CONTAINERS Once rooted, pineapples should be carefully planted in a small pot of peat-free compost mixed with grit in a ratio of 3:1. Alternatively, buy ready-potted plants.

Suspend the pineapple top in a jar of water; change the water regularly.

Empty food tins can be used as pots for young kitchen-scrap plants, but be sure to add drainage holes.

GROW

Keep your pineapples in the sunniest, warmest spot possible, with minimum temperatures of 18°C (64°F). In tropical climates they can be grown outside; in temperate climates they'll either need to be grown indoors or in a heated greenhouse. Make sure sunny windowsills are not draughty or too cold at night. Water the plant regularly during spring and summer and give a liquid feed every two weeks. Plenty of foliage growth means you're more likely to get a large pineapple. Water more sparingly in winter, ensuring the compost doesn't become soggy.

Once the plant is around three years old it should produce a flower spike. If it doesn't, seal the whole plant in a plastic bag with several apples. The ethylene gas released by the apples as they ripen should help to trigger the pineapple plant into flowering.

Baby pineapples are attractive, but not all the varieties that are sold as houseplants are edible (see below).

VARIETIES

If you're rooting your own pineapple plant, the variety of the fruit will be the same as that of the pineapple you buy and will therefore depend on what's available in the shops. The variety should be listed on the fruit label or, if not, on the crate or box from which it was sold.

SMOOTH CAYENNE Most commercially produced pineapples are of this variety. The fruit has firm and juicy flesh and the plant has virtually spineless leaves.

HARVEST

Pineapples can take as long as six months to develop their fruit. Once the fruit has turned orange or yellow and smells sweet, simply cut it off. The plant will then die, but you can root the top of this new pineapple. The plant may also have "pups" – baby plants produced around the base. Cut these off, with a bit of root attached, and pot them up. Pineapples will yield one fruit per plant.

It's worth the wait for the novelty of harvesting a homegrown pineapple.

> **TOP TIP** SOME PINEAPPLES SOLD AS POTTED PLANTS ARE PURELY ORNAMENTAL: THEIR FRUIT IS INEDIBLE. THESE VARIETIES INCLUDE *ANANAS COMOSUS* 'CHAMPACA' AND 'CORONA', AND *A. COMOSUS* VAR. *VARIEGATUS*.

STRAWBERRY *FRAGARIA × ANANASSA*

Homegrown, sun-ripened strawberries are one of the most irresistible fruits to eat and are incredibly easy to grow. Keep your crop productive by replacing plants every three to five years (replace a few a year, in rotation) and each year they'll provide their own baby plants for free.

GROW FROM Potted plants or bare-root runners
HEIGHT AND SPREAD Up to 20 × 30cm (8 × 12in)
HARDINESS Hardy
IDEAL SOIL Moist but well-drained
POSITION/SITE REQUIREMENTS ☼
TIME TO MATURITY 60–90 days

CALENDAR

	SPRING	SUMMER	AUTUMN	WINTER
PLANT	▓▓▓		▓▓	
HARVEST		▓▓▓		

Through the year
By planting different varieties, you can enjoy this quintessential summer fruit from spring through to early autumn.

Plant so that the crown, where roots and shoots meet, is level with the soil.

PLANT

Enhance the flavour and sweetness of your strawberries by planting them in the sunniest spot. Before planting, mix compost or other well-rotted organic matter and a handful of wood ash with the soil to enrich it. Strawberries can be grown from seed, but potted plants or bare-root runners grow more quickly and will bear fruit the summer after planting. Plant potted plants in autumn or spring, and bare-root runners in late summer or spring. Space plants 40–50cm (16–20in) apart.

IN CONTAINERS Place three plants per 40cm- (16in-) diameter pot in peat-free, multipurpose compost. Standard pots are easier to water than "strawberry pots" (with holes in the sides). Watch out for attack by vine weevils (see p.39).

Allowing the fruit to trail over containers will help it to ripen.

GROW

For early spring or autumn harvests, protect plants with cloches or fleece, or grow in a greenhouse. Keep them well-watered, especially if in pots, so that the berries swell evenly. Give a liquid feed fortnightly while fruiting. Deter birds and squirrels; check for slugs (see p.39).

In midsummer, the plants will start producing runners – long stems that bear baby plantlets. If you want more strawberry plants, you can root these runners (see p.48), but only use the first plantlet on the runner for two runners per plant. Cut off the rest. In late winter, tidy up the plants by cutting off dead foliage, and spread a straw mulch around the base.

Straw spread around the base of the plant keeps fruit clean and helps prevent rot, but it can harbour slugs and snails.

Plant breeders introduce new varieties of strawberry with improved flavour, size, or disease resistance every year.

White berries, such as this pineberry, are not as attractive to birds as red berries.

VARIETIES

Strawberry varieties are divided into two groups: perpetual (also known as everbear) and summer-fruiting types. Perpetuals produce a little fruit at a time over a few months; summer-fruiting ripen all at once but are subdivided into those ripening in early, mid-, and late season.

AROMEL Perpetual. Juicy fruit with an intense fragrance.

BOLERO Perpetual. Well-flavoured, reliable berries on vigorous plants.

CAMBRIDGE FAVOURITE Summer-fruiting, mid-season. Heritage variety (see p.23) with excellent flavour. Award of Garden merit (AGM, see p.23).

FRAMBERRY Summer-fruiting, early/mid-season. Small with raspberry-like flavour.

HONEYOYE Summer-fruiting, early season. Makes good jam. AGM.

KORONA Summer-fruiting, early season. Large, ripens fast; good disease resistance.

MALWINA Summer-fruiting, late season. Very dark red, large berries.

MARA DES BOIS Perpetual. Wild strawberry flavour but with bigger berries (bred from *Fragaria vesca, see below*).

MARSHMELLO Summer-fruiting, mid-season. Melt-in-the mouth texture.

PEGASUS Summer-fruiting, mid-season. Good flavour and tolerance of pests and diseases. AGM.

PINEBERRY Summer-fruiting, early/mid-season. A white berry, ripening to light pink, with a hint of pineapple flavour.

ROYAL SOVEREIGN Summer-fruiting, mid-season. Heritage variety with the classic strawberry flavour.

SNOW WHITE Perpetual. White berries.

VIVA ROSA Perpetual. Pink-flowered variety that fruits until the first frost.

HARVEST

Perfectly ripe strawberries are wonderfully aromatic and deep red all over (unlike many of the shop-bought offerings). Check your plants daily for berries that are ready to harvest – do this by snipping or gently breaking the stalk. Make sure the haulm (the green, leafy section on the top of the berry) isn't left behind to rot on the plant. Remove any bad fruit every time you harvest to keep the rest of the crop healthy. A single plant can yield around 450g (1lb) of fruit.

ALSO TRY

Alpine or "wild" strawberries (*Fragaria vesca*) have small berries ideal for garnishing drinks. They crop lightly all summer. Grow the species (*F. vesca*), or try tropical-fruit-flavoured 'White Soul', or 'Candy Floss' (which tastes as its name suggests), or the sweet heritage variety 'Baron Solemacher'.

Wild strawberries will naturalize in steps and paving and grow as ground cover.

GOJI BERRY _LYCIUM BARBARUM_

Goji berries, or wolfberries, widely considered a superfood, are often only available to buy dried, but a single bush provides a plentiful harvest for eating fresh or preserving. The plant forms a large, sprawling, deciduous shrub that copes well with drought and the salty winds of coastal areas.

GROW FROM Potted plants
HEIGHT AND SPREAD Up to 3 × 4m (10 × 13ft)
HARDINESS Hardy
IDEAL SOIL Well-drained
POSITION/SITE REQUIREMENTS ☼
TIME TO MATURITY Two to three years

CALENDAR

	SPRING	SUMMER	AUTUMN	WINTER
PLANT			▓▓	
HARVEST			▓▓	

Through the year
Goji berries can be harvested until the first frosts of autumn.

Long, arching goji berry stems can be cut back to keep the plant smaller, but this will reduce the next harvest.

PLANT

Goji berries are difficult to raise from seed – young plants need protection from frost and the elements. It can therefore be worth investing in an older, hardier potted plant that will grow quickly with little fuss. Ideally, plant against a south- or southwest-facing wall or fence, where their arching, lax stems can be tied up to wires – fix the wires in place before planting. Mix some compost or other well-rotted organic matter into the soil before planting in early autumn. Space at least 3m (10ft) from other plants. Water well and mulch with compost or other well-rotted organic matter after planting.

IN CONTAINERS Plant into a large container (minimum 50cm/20in in diameter) placed at the foot of a wall or fence so the stems can be trained as above. Pots need plenty of drainage holes; fill with a mix of soil-based, peat free compost and grit in a ratio of 3:1.

Goji berry plants are hardy when fully grown, but need extra care while young.

GROW

Tie the growing stems into the wires as needed; in early spring, reduce overcrowding by cutting back some stems to the base. Any branches that touch the ground will form roots there: dig these up and prune or tie back to prevent the plant spreading outwards (rooted sections can be grown into new plants). Protect the plant from harsh conditions during its first winter after planting by covering with horticultural fleece. Mulch in late winter with compost or other well-rotted organic matter mixed with a little wood ash.

Water goji plants regularly for the first two years; established plants are relatively drought-tolerant. Water container-grown plants regularly and give a liquid feed every two weeks.

VARIETIES

There aren't many named cultivars of goji berries, but plant breeders are introducing more as growing the berries becomes increasingly popular. There's still little notable difference in taste between the varieties and the species, or between the varieties themselves. A related species, the black goji (*Lycium ruthenicum*) has black berries that are used in similar ways.

BIG LIFEBERRY Bears slightly more rounded fruit than other varieties; each berry 2cm (¾in) long.

NUMBER 1 LIFEBERRY Heavy yields of large fruits; each berry 2–3cm (¾–1¼in) long.

SWEET LIFEBERRY Fruits are less tart than some other varieties; berries are small, at just 1–2cm (½–¾in) long.

SYNTHIA A more compact plant, especially tolerant of drought and coastal conditions. Sweet fruit is 2–3cm (¾–1¼in) long.

Although unusually small, goji berries are rich in vitamins A and C, whether eaten fresh, cooked, or dried.

HARVEST

Goji berries are edible only when they're fully ripe, which is when they turn a deep red colour. Pick or snip them at the base of the berry to avoid bruising the fruit. Alternatively, gather them by shaking the branches over an old sheet placed on the ground. Eat the berries fresh, or preserve them by drying in a low oven or dehydrator. Yields are variable, but expect between 1–2.5kg (2.2–6lb) per mature plant.

The berries will blacken if handled too roughly, so take care when harvesting.

Dried goji berries can be expensive to buy, but are easily grown and dried at home.

RHUBARB *RHEUM × HYBRIDUM*

Treated as fruit – although it is, technically, a vegetable – this sweet and sour crop works brilliantly stewed, cooked in cakes and bakes, or pickled. One plant will produce a steady supply for a family, but if you want early forced rhubarb you should plant out more.

GROW FROM Potted plants or crowns
HEIGHT AND SPREAD Up to 0.5–1 × 1m (1¾ × 3ft)
HARDINESS Hardy
IDEAL SOIL Moist but well-drained
POSITION/SITE REQUIREMENTS ☼ ◐
TIME TO MATURITY Two years

CALENDAR

	SPRING	SUMMER	AUTUMN	WINTER
PLANT	▨		▨	▨
HARVEST	▨			

Through the year
Rhubarb can be picked from early spring if forced, making it the first fruit of the season.

Keep rhubarb well watered. Don't allow the soil to dry out in hot weather.

Plant rhubarb so that the crown is level with the soil surface.

PLANT

Rhubarb can be grown from seed, but roots and shoots will grow more rapidly from the lumpy, woody "crowns" that can be bought at garden centres. In an open, sunny spot, plant the crown so it is level with the surface of the soil, or slightly higher if the soil is prone to winter wet. Avoid frost pockets and areas that can get waterlogged.

IN CONTAINERS Rhubarb can be grown in pots that are large enough to retain moisture well and to prevent the top-heavy plants from tipping over. Fill the containers with a mix of garden or multipurpose compost and topsoil.

GROW

A rhubarb plant can crop for a decade or more. Apply a layer of compost around the outside of the crown every spring (but don't cover it) and harvest the outer stems regularly to keep the plant compact. Cut off flower stalks at the base if they appear. Rhubarb isn't usually troubled by pests and diseases. In autumn, pull away all the old stems and leaves; the crown needs exposure to cold in order to stimulate it into growth next spring.

To force rhubarb, cover the crown securely with an upturned large pot or bin in late winter. It must exclude all light from the plant. Keep it covered until the stems are a harvestable size, then pick them and remove the cover.

Roasting, baking, and pickling best preserve the pink colouring of rhubarb stems.

VARIETIES

For forced rhubarb, choose early or vigorous varieties. Stems of mid-season rhubarbs, harvested from early spring to early summer, start off red but can fade to green. Rhubarb is sometimes sold as *Rheum cultorum*, a synonym for *R. × hybridum*.

BAKERS ALL SEASON If planted in a frost-free spot this variety can grow and be harvested all year round.

CAWOOD DELIGHT Mid-season. Dark red stems keep their texture and colour well when cooked. Good choice for small spaces.

FULTON'S STRAWBERRY SURPRISE Mid-season. Grows well in containers and has excellent flavour. Award of Garden Merit (AGM, see p.23).

RASPBERRY RED Mid-season. Heritage variety with sweet, deep red stems. AGM.

STEIN'S CHAMPAGNE Mid-season. Slender stems are red along whole length.

TIMPERLEY EARLY Early season. Suitable for forcing. Slender stems that don't get woody. AGM.

VICTORIA Can be forced or harvested mid- to late season. Vigorous variety from the 1850s. Plants have greenish stems.

HARVEST

Rhubarb plants need time to gather strength, so don't harvest any stems in the first year and take only a few stems in the following year. After this you can harvest every spring, but avoid pulling up more than a third of the stems at once. Stop picking in late summer to allow the plant to build up the energy reserves it requires to see it through the winter.

Harvest forced stems all at once when they're ready, then leave the plant to recover for the rest of the year.

Don't cut rhubarb stems – they're ready when they detach themselves easily. Take a firm grip just above the base of the stem and pull. Discard the leaves, which are poisonous.

Flower stalks reduce the vigour of the plants – cut them off as they appear.

Pull rhubarb stems away from the base rather than cutting them.

BLACKCURRANT *RIBES NIGRUM*

Blackcurrants are probably the easiest to grow of all the soft fruits and very productive: a bush for the same price as a few punnets of fresh currants offers a great return on your investment. The fruits are rich in vitamin C and antioxidant anthocyanins, which give them their purple colour.

GROW FROM Potted or bare-root plants
HEIGHT AND SPREAD Up to 2 × 2m (6ft 6in × 6ft 6in)
HARDINESS Hardy
IDEAL SOIL Moist but well-drained
POSITION/SITE REQUIREMENTS ☼ ☼
TIME TO MATURITY Four years

CALENDAR

	SPRING	SUMMER	AUTUMN	WINTER
PLANT				
HARVEST				

Through the year
Plant bushes of different varieties to spread the harvest between summer and early autumn.

Plant blackcurrants in large containers; add drainage holes if necessary.

Use a piece of wood to help ensure the plant's crown is 5cm (2in) below soil level.

Firm the soil in well around the plant's roots before watering.

PLANT

Blackcurrants are especially prone to viruses, so make sure your plants are certified as disease-free stock. You'll only need to access them twice a year (for harvesting and pruning), so they can be tucked at the back of a flower border or grown in a forest garden setting (see p.16) as they'll tolerate dappled shade. Allow 1.5–2m (5ft–6ft 6in) between plants, 1.2m (4ft) for compact varieties.

Ideally, plant blackcurrant bushes in autumn, when the soil is still warm; if you're planting in winter, avoid frozen ground. Before planting, enrich the soil by mixing in compost or some other type of organic matter. Then dig a hole deep enough to accommodate the roots: to stimulate new shoots, bury the crown of the bush (where the roots turn to shoots) 5cm (2in) below soil level. Add a layer of mulch around the base of the bush after planting and watering in.

IN CONTAINERS Blackcurrants can be grown in large containers or raised beds, in a soil-based, peat-free compost. The more upright, compact varieties, such as 'Ben Nevis' and 'Ben Connan', are better suited to containers. Prune spreading branches (see p.115). Repot every other year (even if returning the plant to the same pot); knock off loose compost and replace it with fresh. Protect flowering blackcurrants from frost by moving the pots into a greenhouse if you have one.

VARIETIES

All blackcurrants are self-pollinating and high-yielding, so you need grow only one plant in your garden, if that's all you want. The 'Ben' varieties have been bred for good disease resistance and hardiness against late frosts. Earlier-flowering types may have poorer fruit set if there are cold winds or frosts in early spring.

BALDWIN Mid-season. Heritage variety (see p.23); tart, flavoursome, thick skinned.

BEN CONNAN Early to mid-season. Bushy, compact plants and high yields. Award of Garden Merit (AGM, see p.23).

BEN LOMOND Mid- to late season. Good for colder areas with reliable heavy crops.

BEN NEVIS Mid- to late season. Upright, vigorous plants that do well on poor soils; flavoursome, juicy currants.

BEN SAREK Early season. Relatively compact plants but with spreading branches; tart fruit.

BIG BEN Early season. Bears currants double the size of other varieties and therefore often juicer. AGM.

TITANIA Mid-season. High yields of juicy currants with excellent flavour.

WELLINGTON XXX Early to mid-season. A heritage variety. High yields of large fruit especially good for jam.

The berries of 'Big Ben' are large and therefore juicier than other varieties.

HARVEST

Wait to pick blackcurrants for a few days after they've turned a deep purple/black, testing them daily, so that they develop their sweetness. Pick all the ripe berries, leaving those that aren't fully coloured – these will provide a second crop in a few weeks' time.

The leaves also have a delicious blackcurrant flavour; infuse them into creamy desserts or sugar syrups (for cakes or sorbet), or steep them in hot water for a herbal tea.

Each bush will produce around 4.5kg (10lb) of fruit once mature, with smaller harvests in the preceding years. Late frosts can reduce yields.

Blackcurrants give a heavy first picking, then a smaller secondary one if the unripe fruits are left on the plant to develop fully.

Blackcurrant plants are hardy, but their tiny flowers may need protection from frost and wind (see also above).

GROW

Container-grown blackcurrants should be watered regularly and given a liquid feed fortnightly. Water blackcurrants in the ground during dry spells. Prune bushes in winter (see p.115), then mulch with compost or well-rotted organic matter mixed with a little wood ash.

Look out for extra-swollen buds in early spring, which are a sign of a big bud mite infection; the mites also transmit diseases, so destroy infected plants. The bushes may also be affected by aphids, but will tolerate low-level infestations well (see pp.38–39).

REDCURRANT *RIBES RUBRUM*

Redcurrants give good yields of delicious, tart fruit on low-maintenance plants. They're easily adapted to any garden situation: grow as single- or multi-stem cordons or fans against a fence, wall, or freestanding posts and wires; or in beds, borders, or pots as bushes or cordons.

GROW FROM Potted or bare-root plants
HEIGHT AND SPREAD Up to 1 × 1m (4 × 4ft)
HARDINESS Hardy
IDEAL SOIL Moist but well-drained
POSITION/SITE REQUIREMENTS ☼ ☼
TIME TO MATURITY Six years

CALENDAR

	SPRING	SUMMER	AUTUMN	WINTER
PLANT				
HARVEST				

Through the year
Currants come into leaf early in spring, providing welcome greenery, and trained forms add winter structure to the garden.

PLANT

If you want to train currant plants on a wall, fence, or wires between posts (see *pp.116–117*), put in all your wires before planting. Redcurrants will tolerate almost any soil and situation, including dappled shade and north-facing walls (although they'll fruit around two weeks later than the same variety grown in full sun). They dislike waterlogged soil though, so if necessary improve the soil's drainage by incorporating some gravel where you want to plant them. Mix in some compost or other well-rotted organic matter before planting to enrich the soil.

Plant in autumn ideally, or winter if the ground isn't frozen, then water and mulch with compost or other organic matter. Allow 1.2m (4ft) between bushes, 45cm (18in) between cordons, and 1.5–2m (5–6ft 6in) between fans.

IN CONTAINERS Grow redcurrants in large freestanding containers as bushes or single cordons; you can also plant them in raised beds, pots, or troughs at the base of a wall or fence and train as above. Plants grown as cordons in pots will need the support of a stake. Use a peat-free, soil-based compost. If possible, move the containers in spring to protect the flowering plants from frost and cold winds.

Currants such as these, grown as cordons against a shed wall, are ideal for making the most of out-of-the way spaces.

GROW

Pot-grown redcurrants need regular watering and a fortnightly liquid feed in spring and summer. Plants in the ground need watering while establishing and during dry spells, and an annual mulch of compost or other well-rotted organic matter, mixed with a handful of wood ash, in late winter.

Prune bushes while dormant (see *p.114*); trained forms need winter and summer pruning (see *p.116*). Check the ties of trained forms periodically, tying in and loosening as necessary.

If necessary, use netting to deter birds, as on these 'Red Lake' currants, but fasten it well to avoid ensnaring them.

VARIETIES

All redcurrants are self-pollinating. The fruit is generally tart rather than sweet, and all varieties are suitable for eating fresh or preserving as jellies and jams. The currants also freeze well. Choice of variety is determined more by harvest period and size of individual currants than by flavour, as all varieties are similar in taste.

JONKHEER VAN TETS Early season. Heavy yields of large currants. Award of Garden Merit (AGM, see p.23).

JUNIFER Early season. Long strigs of currants; plant displays good resistance to disease.

LAXTON'S NUMBER ONE Mid- to late season. A reliable and vigorous heritage variety (see p.23) with long strigs.

RED LAKE Mid- to late-season. Heavy crops of currants borne on long strigs. AGM.

ROVADA Late season. Large, juicy berries and reliable heavy yields. AGM.

STANZA Mid-season. Reliable yields of good-flavoured currants. AGM.

'Jonkheer van Tets' reliably provides prolific yields of currants.

HARVEST

Redcurrants are ripe once they've turned fully red, but leave them for a few more days before picking to develop their sweetness. The currants grow on little pendulous strings of stalks called strigs or trusses, which hang from the woody stem. It's easiest to pick off the whole strig when harvesting and then detach the fruit from the stalk in the kitchen: simply run the strig through the tines of a fork to quickly ping off the currants.

You'll get a decent harvest from the plant's second year, but redcurrants aren't fully productive until they're around six years old. On mature plants, expect 4–5kg (9–11lb) of fruit per bush plant, or 1kg (2lb) per cordon.

ALSO TRY

Whitecurrants are pale varieties of redcurrants, and will get pinker the longer they're left to ripen. Heritage variety 'Versailles Blanche' (syn. 'White Versailles') was first introduced in 1843 (see p.23) – it produces flavoursome but seedy currants. 'White Grape', has strong upright growth and an Award of Garden Merit (see p.23). Delicately fragranced 'Gloire de Sablon' is a pink currant, a cross between a red- and a whitecurrant.

Whitecurrants are less obvious to birds than their red relations and therefore often go untroubled.

GOOSEBERRY *RIBES UVA-CRISPA*

If you can find gooseberries in the shops, they're usually bullet-hard and fit only for cooking. Growing your own means you can enjoy ripe berries fresh from the garden, and still have plenty for jam and desserts. Also known as grozets or goosegogs, gooseberries can be green, yellow, or red.

GROW FROM Potted or bare-root plants
HEIGHT AND SPREAD Up to 1.5 × 1.5m (5 × 5ft)
HARDINESS Hardy
IDEAL SOIL Moist but well-drained
POSITION/SITE REQUIREMENTS ☼ ◐ ☀
TIME TO MATURITY Four years

CALENDAR

	SPRING	SUMMER	AUTUMN	WINTER
PLANT				
HARVEST				

Through the year
It's worth braving the thorns for the gooseberry harvest, the earliest soft fruits of the year.

Plant gooseberries so the root flare (where roots become stems) is at soil level.

Use pots to squeeze gooseberry plants into the smallest of spaces.

PLANT

Gooseberries grow happily in most soils and situations, including dappled shade, but avoid frost pockets. Training them as cordons or fans (see p.116, put wires in place before planting) makes picking easier. Plant in autumn or winter, but not into frozen ground. Allow 1.2m (4ft) between bushes, 30cm (12in) between single stem cordons, and 1.5m (5ft) between fans. Then water and mulch with compost or other organic matter.

IN CONTAINERS Grow in large freestanding containers as bushes or single cordons (supported with a stake), or put the pot at the base of a wall or fence to train plants on wires (see right). Plant in peat-free, soil-based compost.

GROW

Water gooseberries regularly to avoid the fruit splitting and mildew taking hold (see p.41); give pot-grown gooseberries a fortnightly liquid feed in spring and summer. Mulch every late winter with compost or other well-rotted organic matter, mixed with a handful of wood ash. Protect the flowers from frosts by covering with horticultural fleece. Prune gooseberry bushes in winter (see p.114) and also in summer for trained forms (see p.117). Regularly check and loosen ties as needed, and tie in new shoots.

This 'Invicta' gooseberry plant has been trained on a fence for easier picking.

VARIETIES

Of the 3,000 or so gooseberry varieties recorded since the 1700s, 150 survive. All are self-pollinating. Some heritage varieties are delicious, but new cultivars have fewer thorns or better resistance to problems such as mildew (see p.41).

CARELESS Early season. Culinary. Reliable heritage variety; green berries. Award of Garden Merit (AGM, see p.23).

GREENFINCH Mid-season. Culinary/ dessert. Green berries on compact plants with good disease resistance. AGM.

HINNONMÄKI RÖD (syn. 'Hinnonmaki Red') Mid-season. Dessert. Sweet, red, aromatic berries; hardy and reliable.

INVICTA Mid- to late season. Culinary. Heavy crops of green berries on mildew-resistant plants. AGM.

LEVELLER Mid-season. Dessert. Heritage variety with reliably heavy crops of superbly flavoured yellow berries. AGM.

MARTLET Late season. Culinary/dessert. High yields of small, red, smooth-skinned berries borne on mildew-resistant plants.

PAX Mid-season. Culinary/dessert. Large berries on plants bred for fewer thorns.

WHINHAM'S INDUSTRY Mid-season. Culinary/dessert. Heritage variety. Cook when green; ripen to red for eating. AGM.

'Pax' forms an upright bush with almost spineless branches and flavoursome red berries.

ALSO TRY

The Worcesterberry (*Ribes divaricatum*) is a self-pollinating, thorny North American relation of the gooseberry. The fruit is similar to a gooseberry, but with hints of blackcurrant; ripens mid- to late summer. Grow and prune as for gooseberry; allow 1.8m (6ft) between bushes. Jostaberry (*R. × culverwellii*) is closer to the blackcurrant, but still with looks and flavour attributable to the gooseberry. Plants reach 1.5m (5ft) in height and spread, and are tolerant of most soils and situations: grow and prune them as you would blackcurrants (see p.100). Harvest in early summer.

Jostaberries taste and look like a cross between a gooseberry and blackcurrant.

HARVEST

Gooseberries are ripe and ready to pick when they're just slightly soft. Pick them with a bit of stalk attached to avoid tearing the skin of the fruit, then top and tail them in the kitchen (nick off the stalk from one end and the old flower from the other).

The berries can be for culinary use, eating fresh (dessert), or both. If choosing the latter, "thin" (pick) alternate fruits on each branch: the first harvest of gooseberries will be for cooking, leaving the rest to ripen into larger berries for eating fresh. Gooseberry yields can be between 3.5–4.5kg (7–10lb) from a mature bush, or 1kg (2lb 4oz) from a single cordon.

CHILEAN GUAVA *UGNI MOLINAE*

The small, aromatic berries of this plant are more similar in taste to raspberries, blackcurrants, and blueberries than to the ball-sized fruit of a tropical guava. Chilean guavas grow on evergreen shrubs with dense foliage and pretty flowers – the plants make ideal hedging to edge paths or fruit beds.

GROW FROM Potted plants
HEIGHT AND SPREAD Up to 1.5 × 0.5–1m (5 × 1ft 9in–3ft)
HARDINESS Half-hardy/Hardy
IDEAL SOIL Moist but well-drained
POSITION/SITE REQUIREMENTS ☀
TIME TO MATURITY Three to five years

CALENDAR

	SPRING	SUMMER	AUTUMN	WINTER
PLANT			▓▓	
HARVEST			▓▓	

Through the year
The late ripening of Chilean guavas in the depths of autumn brings welcome freshness to the end of the fruit season.

Chilean guava's nodding, bell-shaped flowers add another season of interest to this attractive plant.

Lay out all your plants before starting to plant them as a hedge to ensure equal and sufficient spacings.

PLANT

In their wild habitat, Chilean guavas grow in dappled shade on the edges of woodlands. This makes them ideal plants for a forest garden setting, or for planting among ornamentals in borders, although they also enjoy a sunny spot. Planting them near the house or a path means you'll be able to more easily enjoy their wonderful fragrance in autumn. They prefer a sheltered spot, not exposed to cold winds or frost.

Ideally, plant in autumn; plant all year round if the ground isn't frozen. Water well while the plants are establishing. Allow up to 1m (3ft) between the plants, or 50cm (20in) if planting as a hedge.

IN CONTAINERS Chilean guavas are well-suited to growing in containers. In colder areas, containers are preferable to the open ground because the plants can be more easily protected from frosts and exposure to strong winds, either by covering with fleece or by bringing them under the temporay cover of a greenhouse or some other structure.

Pots must be big enough to ensure plants have plenty of root space and compost won't dry out too quickly, and to prevent top-heavy potted plants from toppling over: aim for a 50cm (20in) pot for a mature plant (1.5m/5ft tall). Use peat-free, soil-based compost, repotting into a larger container every other year.

GROW

These are generally low-maintenance plants, untroubled by pests and diseases. Keep them well-watered in dry spells, and water those in containers regularly. Give potted plants a liquid feed every two weeks from spring to summer. Mulch annually in late winter with compost or well-rotted organic matter, mixed with a handful of wood ash.

If possible, protect plants, especially young ones, from frost. Trim branches in early spring, taking off frost-damaged tips and shaping the plants, if required: for example, to form a rounded hedge.

VARIETIES

Chilean guavas are most often sold as the generic species, although the varieties below may be available. They sometimes appear under the synonym *Myrtus ugni* and are also known as strawberry myrtle. In cooler climates, if planting as a freestanding bush or tall hedge, it's worth looking out for the variety 'Ka-Pow'. All Chilean guavas are self-pollinating.

BUTTERBALL This variety has yellow leaves and pinkish stems on the young foliage; the berries are the same as the species.

FLAMBEAU The berries taste no different from the species, but foliage is variegated.

KA-POW (syn. 'Yanpow') A British-bred variety that produces bigger berries than the species, starts fruiting at a younger age, and is more tolerant of cold weather. Reaches up to 1.5m (5ft) tall.

The ripening berries on the variety 'Ka-Pow' start off pale in colour, and then ripen to a wonderful claret shade of red.

HARVEST

Chilean guavas are ready to harvest in late autumn and early winter. Leave them on the plant until they're fully ripe – if necessary, check them on a daily basis to make sure. They'll be a deep red-purple colour, and soft, sweet, and flavoursome when ripe.

These fruits are delicious eaten fresh, but they can also be preserved as jam or used in baking – simply substitute them in any recipes that call for blueberries. The berries may not all ripen at the same time – if you want a significant quantity of the fruit to preserve, then just freeze the earlier harvests until you have enough.

Yields from a mature bush can be up to 1kg (2lb), or sometimes even more in a good year. Hedging plants, which are individually smaller, will have lower yields of fruit per bush.

As the fruit gradually ripens, it gives off a magnificent aroma, reminiscent of strawberries.

Fuchsia **'Riccartonii'** is one of the hardiest fuchsias.

ALSO TRY

Fuchsias are grown ornamentally, but the flowers and berries are edible. They need a sheltered spot to yield berries and are unlikely to produce more than 0.5kg (1lb) per plant, but there are more prolific varieties such as 'Fuchsiaberry'. 'Globosa' also gives a relatively heavy harvest and *F. magellanica* is hardy.

BLUEBERRY *VACCINIUM CORYMBOSUM*

Blueberry breeding for commercial use has created fruits that are larger and sweeter but can be watery and less tasty. The blue flesh (coloured with antioxidant anthocyanins) has almost disappeared in shop-bought blueberries. But growing your own means you can eat delicious, healthy berries all summer.

GROW FROM Potted or bare-root plants
HEIGHT AND SPREAD Up to 1.5–2 × 1.5m (5ft–6ft 6in × 5ft)
HARDINESS Hardy
IDEAL SOIL Moist but well-drained; acidic
POSITION/SITE REQUIREMENTS ☼ ☽
TIME TO MATURITY Six years

CALENDAR

	SPRING	SUMMER	AUTUMN	WINTER
PLANT				
HARVEST				

Through the year
Attractive all year: from white spring blooms, glaucous summer foliage and fruit to bright autumn leaves and winter stems.

Blueberry foliage should be a deep grey-green colour: yellowing leaves may mean the plant is struggling for nutrients because the soil pH is too high (alkaline).

PLANT

Blueberries need a sunny spot or dappled shade and an acidic soil of between pH 4 and pH 5.5 (see pp.14–15): if your soil isn't suitable, it's better to grow blueberries in raised beds or containers.

Space your plants 1.5m (5ft) apart. Plant them in autumn or winter (but not into frozen ground), then water and mulch with compost or other organic matter afterwards.

IN CONTAINERS Containers are perfect for blueberries. Plant them into a mix of peat-free, ericaceous (lime-free) compost and composted bark in a ratio of 3:1, potting on the plant into a bigger container every other year as it grows, finishing with a 50cm (20in) pot for a 1.5m- (5ft-) tall bush. This will ensure they're not top heavy, and give the plants plenty of root space in compost that won't dry out too quickly.

If growing blueberries in raised beds, these should also be filled with a 3:1 mix of peat-free ericaceous compost and composted bark.

GROW

Water blueberries in pots regularly, and those in the ground while establishing and during dry spells, especially when flowering and fruiting. Use rainwater if possible (installing a water butt is a good idea, see p.37) as tap water, particularly in hard-water areas, can raise soil pH. Once a month during spring and summer add a liquid feed to the water for plants in pots. In late winter every year, spread a mulch around the base of the bushes – either of compost, composted bark, or pine duff (old pine-tree needles) – to help keep soil acidic. Avoid composted horse manure, which can make the soil more alkaline.

Blueberries are relatively untroubled by pests and diseases, and only need minimal pruning in winter once they're fully grown (see p.115).

Blueberries are ideal potted plants for a sunny courtyard; this one is the variety 'Patriot' (see *opposite*).

VARIETIES

Blueberries are only partially self-pollinating. A single bush will bear fruit, but two or more will be better pollinated and therefore give bigger harvests, unless your neighbours also have blueberry plants. Ideally, choose three varieties, one each from early, mid- and late season; they all flower at around the same time, but their ripening will be staggered across the season.

BLUECROP Mid-season. Large fruits (it's a commercial favourite) with a good flavour.

DUKE Early season. Large bushes have late flowers that often avoid frosts, leading to high yields. Award of Garden Merit (AGM, see p.23).

HERBERT Mid- to late season. Berries are very large with a good acid/sugar balance. High yields and a spreading habit.

PATRIOT Early season. Reliable plants have a good tolerance of cold weather and heavy, damp soil. Good, complex flavour.

PINK LEMONADE Mid-season. A pink berry with very sweet, floral flavour but low yields.

RUBEL Mid-season. A heritage variety (see p.23); the small berries have shown in tests to have very high anthocyanin levels.

SPARTAN Early season. Berries have an excellent flavour; plants display good bronze/gold autumn colour; AGM.

Blueberries don't ripen all at once, but harvest regularly over a long period.

HARVEST

When the berry is deep blue in colour and detaches easily from the stalk it's ready to harvest. Blueberries aren't particularly prone to gluts, but they do freeze well should you want to preserve the harvest. Younger blueberry plants give smaller harvests; a mature bush will yield between 2.25 and 5kg (5 and 11lb) of fruit.

Ripe, juicy blueberries can be carefully "tickled" off the stalks into a waiting container.

ALSO TRY

Darrow's blueberry is not a true blueberry but a related species, *Vaccinium darrowi*, sometimes sold as blueberry 'Darrow'. It's an excellent fruiting and ornamental evergreen shrub with pretty pink blossom in spring. The plants are compact (1.5m/5ft in height and spread) and bear large and tasty blueberry-like fruits in summer.

CRANBERRY *VACCINIUM MACROCARPON*

Cranberries are evergreen, low-growing, low-maintenance plants, producing bright red berries that are fast becoming a staple of UK seasonal dining and baking. They're ready to harvest in late autumn and store well, so you can look forward to fresh cranberries with your festive winter dinners.

GROW FROM Potted plants
HEIGHT AND SPREAD Up to 50 × 50cm (20 × 20in)
HARDINESS Hardy
IDEAL SOIL Moist, acidic
POSITION/SITE REQUIREMENTS ☼ ☀
TIME TO MATURITY Two to three years

CALENDAR

	SPRING	SUMMER	AUTUMN	WINTER
PLANT	▓▓		▓▓▓	▓▓
HARVEST			▓▓	

Through the year
Cranberry plants are impressively productive: the plants will keep going for up to 100 years.

Containers such as this old sink can be sunk into the ground, rather than remaining freestanding, to create a cranberry bed.

You can plant potted cranberries at any time of the year, but between autumn and spring is best.

PLANT

Cranberries need soil that is moist to the point of boggy and with a pH of 4–5.5 (see p.15). If your garden isn't suitable, you can either plant into a raised bed or container, or create a cranberry bed.

To create a cranberry bed within the garden, choose a site in full sun or dappled shade and dig out the soil to a depth of the blade of the spade and as long and wide as needed to accommodate all your plants when spaced 30cm (12in) apart. Line the

pit with plastic pond liner sheeting, puncturing it in a few places to allow some drainage. Aim to keep the soil very moist but not waterlogged. Fill the pit with peat-free ericaceous (lime-free) compost, plant, and water well.

IN CONTAINERS Containers for cranberries don't need to be deep but they do need to be wide. They also shouldn't have too many drainage holes: old sinks and troughs are ideal. You can also grow cranberries in hanging baskets lined with plastic sheeting; puncture it in just a couple of places for drainage to help keep the compost moist.

Fill pots with peat-free ericaceous compost – plant the cranberries into their final, biggest, pot immediately rather than moving them up a size each year as they dislike disturbance.

GROW

Water as required to keep the soil moist. Give plants in containers a liquid feed once a month during spring and summer. If the soil becomes too alkaline, the leaves will start yellowing (see also p.112). Mulch in late winter, using compost, composted bark, or pine duff (old pine needles); avoid products containing horse manure, which will raise the soil pH (reduce the acidity). Jiggle the stems so that the compost settles underneath the foliage.

Plants need minimal pruning in spring and autumn (see p.115).

Using rainwater from a water butt helps the environment and is less likely to raise the soil pH than tap water, helping to keep your cranberry plants healthy.

The longer the fruit is left to ripen on the plant, the sweeter the berries will be.

HARVEST

Cranberries are ready when deep red in colour; harvest before the first frosts. For easier picking, wait until most of the fruits are ripe and harvest them all at once, either by combing or tickling the berries off the stems, or by picking one by one. They store well in the fridge, or you can freeze them. Expect yields of 0.5–1kg (1–2lb) per plant, once mature.

VARIETIES

To ensure good pollination, and therefore high yields of fruit, it's best to grow at least two cranberry plants of the same variety. There isn't a large choice of available varieties; breeding has been focused on producing plants with larger berries rather than any discernible differences in flavour. Note that the plant commonly known as "Highbush cranberry" (*Viburnum trilobum*) is not a true cranberry.

EARLY BLACK Fruit ripens early (in late summer); very dark red berries are relatively sweet.

FRANKLIN Heavy crops of medium-sized berries that are low in juice; ideal for sauce-making.

MCFARLIN Large, dark red fruits with a waxy bloom ripen late on this US heritage variety (see p.23).

PILGRIM A modern variety with juicy berries and some resistance to blossom damage by frost.

Cranberry 'Pilgrim' has large berries that ripen from white to a deep red.

TOP TIP MOST DRIED CRANBERRIES ARE SOLD SWEETENED WITH ADDED SUGAR. MAKE YOUR OWN HEALTHIER, UNSWEETENED VERSION BY DRYING FRUITS IN A DEHYDRATOR OR AN OVEN SET TO A LOW TEMPERATURE.

LINGONBERRY *VACCINIUM VITIS-IDAEA*

Also known as mountain berries and cowberries, lingonberries have a tart edge to their sweetness that's great in desserts and savoury dishes. These are low-maintenance, compact shrubs; plant dwarf varieties as edible ground cover under and around other shrubs that need acidic conditions, such as blueberries.

GROW FROM Potted plants
HEIGHT AND SPREAD Up to 75 × 75cm (30 × 30in)
HARDINESS Hardy
IDEAL SOIL Moist but well-drained; acidic
POSITION/SITE REQUIREMENTS ☼ ☼
TIME TO MATURITY Three years

CALENDAR

	SPRING	SUMMER	AUTUMN	WINTER
PLANT				
HARVEST				

Through the year
Lingonberries take a few months to ripen, making an attractive display as they do so.

GROW

Water young plants and those in pots (including sunken pots) regularly, and older plants during dry spells. If possible, use rainwater rather than tap water, which can raise the soil pH, especially in hard-water areas. Apply a liquid feed to container-grown plants every month during spring and summer. Mulch annually in late winter with compost (homemade or peat-free, ericaceous), composted bark, or pine duff, not products using horse manure as these can make the soil more alkaline.

Lingonberries need little to no pruning (see p.115). In very harsh winters, the plants can lose some or all of their leaves, but they will grow anew in spring.

Yellowing of the leaves, known as "chlorosis", can be a result of the soil becoming too alkaline.

PLANT

Although lingonberries prefer a sunny spot, they'll happily tolerate dappled shade. They require an acidic soil of between pH 4 and 5.5 (see p.15) and are better grown in containers if your garden soil isn't right. The containers can be sunk into the ground if you want to use the plants in a forest garden setting or as ground cover to help suppress weeds (see right).

Plant in autumn, ideally, or in winter or spring, allowing 50cm (20in) between plants, 30cm (12in) for dwarf varieties. Water well and apply a mulch of peat-free ericaceous (lime-free) or homemade compost, composted bark, or pine duff (old pine needles) after planting.

IN CONTAINERS Fill pots and beds with a 3:1 mix of peat-free ericaceous compost and composted bark. To sink pots: plant into a large pot (it's not easy to repot sunken pots), dig a hole, and fit in the pot.

A mulch of pine duff around the base of plants helps to keep the soil acidic.

VARIETIES

Plant breeding is bringing new varieties of lingonberries onto the market, although they differ more in growth habit, reliability in fruiting, and berry size than in taste. They also differ little from the species, which will be relatively less expensive than the named varieties – so unless you need plants of a particular size, they might not be worth the extra outlay. Lingonberries are self-pollinating, but having more plants is likely to result in better yields.

FIREBALLS Robust, prolific plants reach around 75cm (30in) high.

IDA Dwarf plants of only 20cm (8in) high are good for ground cover or containers.

KORALLE GROUP (syn. 'Koralle') Reliable plants reach around 40cm (16in) high. Berries are early to mature, medium-sized, and tart. Award of Garden Merit (AGM, see p.23).

RED PEARL Selected from a wild strain, these plants reach 40cm (16in) high and have small, flavourful berries.

Lingonberry plants can be flowering and developing fruit at the same time, such as this Lingonberry Koralle Group.

HARVEST

Lingonberries flower twice within the growing season, so it's possible to have a double harvest: one in mid- to late summer and the other in early to mid-autumn. However, late frosts can kill the early blossom and deny you the first harvest. The fruits develop quickly after flowering but take a long time to ripen; they should be an intense red colour all over and sufficiently sweet. Test them regularly to be sure that the sweet/tart balance is to your liking before harvesting. Yields aren't large, only 0.5–1kg (1–2lb) per plant, but a ground cover group of several plants will multiply that harvest easily.

ALSO TRY

Like the lingonberry, cloudberries (*Rubus chamaemorus*) are a popular foraged delicacy in Scandinavian countries. In Canada and northern USA they're known as baked-apple berries. The cloudberry is a member of the same family as raspberries and resembles a miniature blackberry in its growth habit. It favours boggy acidic soil and very cold conditions, and needs these to fruit well.

In some regions, such as the UK, a prevalence of male plants means fruit is a rarity in the wild.

PRUNING FRUIT BUSHES

Fruit bushes need only minimal pruning to keep them healthy and neat, taking very little investment of your time each year. In return, you'll be rewarded with productive plants that provide you with bowlfuls of berries every summer. It's hard to go wrong pruning these simple shapes, but it's worth familiarizing yourself with the pruning basics before getting started (see pp.34–35).

These gooseberries are growing on the darker, older wood, with the paler-stemmed new growth shooting from above them.

PRUNING REDCURRANTS AND GOOSEBERRIES

Freestanding redcurrant and gooseberry bushes are pruned in exactly the same way (see pp.116–117 for pruning trained redcurrants and gooseberries). There are various pruning methods, but the most straightforward, known as "renewal pruning", is described below. Pruning is carried out in early spring and follows the same principles whatever the age of the plant. Removing some branches each year stimulates the growth of new stems, thereby renewing the bush. The aim is to allow air and light into the bush by creating a balanced framework of healthy stems shaped like a vase (see right) – an open centre with stems evenly spaced around the edge. This makes it easy to pick the fruit.

Ideally, the stems should arise from a single central trunk or "leg" of around 10–20cm (4–8in) tall. Remove broken, dead, or diseased wood (see p.84).

> **TOP TIP** WHILE PRUNING, TAKE THE OPPORTUNITY TO REMOVE ANY WEEDS THAT HAVE SEEDED AROUND THE BASE OF THE PLANTS – THEY'RE EASIER TO SEE AND TAKE OUT NOW THAN WHEN THE PLANT IS IN LEAF.

Cut out up to a third of older stems to leave a strong, balanced structure; also cut out any stems growing from the base.

Remove any branches crossing the centre or crowding other, better-placed shoots, prioritizing older wood. Don't take out more than around a third of the branches in a single year, including dead ones. Any stems growing out from the base should be cut back to the trunk to keep the leg clear (see above).

In summer, a second pruning will help maintain good air circulation around the bush. Simply shorten all new shoots to two or three leaves from the main stem.

Summer prune your redcurrant bush by taking off the new growth back to two or three leaves from the main stem.

Make clean cuts with sharp tools to avoid introducing disease to the plant.

PRUNING OTHER BUSHES

Trim the ends of trailing and spreading cranberry stems after harvesting in autumn (see p.111). In spring, prune to prevent overcrowding by cutting out excess stems at soil level, leaving a single layer of stems over the ground.

Lingonberry (see pp.112–113) and Goji berry (see p.96–97) bushes only need dead or broken stems removing in early spring. Goji berry bushes can also have some older stems removed to reduce overcrowding – cut them out at ground level. Pineapples, strawberries, and rhubarb need no pruning (see pp.92–93, pp.94–95, and pp.98–99). Prune blueberries in the same way as blackcurrants (see left), but remove only one or two of the worst-placed, older stems each year after the bush has reached maturity (about six years).

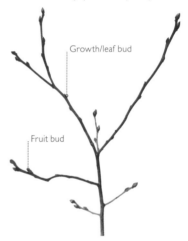

Growth/leaf bud

Fruit bud

When pruning blueberries, remove stems that have proportionally fewer fruit buds as these will be the least productive.

NEED TO KNOW
- All fruit bushes bear fruit on branches of at least one year old.
- Fruit buds are larger than leaf buds (see above).
- Older wood is darker and harder than pale, bendier new growth.

PRUNING BLACKCURRANTS

Blackcurrant pruning promotes the growth of new shoots from the base, keeping the plant's growth continually refreshed, and promoting ripening by allowing light and air to penetrate the centre of the plant. Prune the bush once a year, in the dormant season, cutting stems back to ground level. Prioritize the removal of dead, damaged, or diseased stems; drooping branches; and those that are weak or crowded. Aim to remove a total of around a third of the branches to create a balanced shape, with the remaining stems evenly spread around the bush and with a mix of old and new stems.

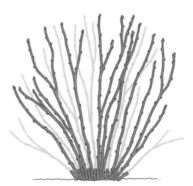

Cut blackcurrant stems back to within 2cm (1in) of the soil level if possible.

TRAINING AND PRUNING FRUIT BUSHES

Gooseberry and currant bushes (red-, white-, and pinkcurrants) can be easily trained as cordons against walls, fences, or onto wires that have been stretched betweenmtwo freestanding posts. This straightforward way of growing the bushes is economical on space, can provide them with some shelter to help the fruit ripen, and also means that the fruit is easier to pick.

Make the most of a sunny wall to train a gooseberry or redcurrant plant.

SINGLE CORDONS

Cordons are trained upright forms of a bush or tree, restricted to a columnar shape. Put training wires in first, 60cm (24in) apart vertically, with a cane tied in to support the upright growth (as for fruit trees, see p.86). In early spring, plant a year-old plant (or several, 30–40cm/12–16in apart). Cut the main stem back by half and all the sideshoots back to one bud from the main stem.

Prune after planting in winter, unless the nursery has already done so.

Growing fruit cordons allows for many different varieties to be grown in one space.

MULTIPLE CORDONS

Currants and gooseberries can be trained as fans, but they often look messy, with lots of canes secured to the wires, and are quite complicated to establish. A better option is to go for a multiple cordon: upright stems trained into a two- or three-pronged fork shape ("double" or "multiple" cordons).

To start forming a double cordon, plant as above, then find a point just above the height of the lowest wire where there are two buds nearly opposite each other. Prune to just above the topmost bud. Trim any other shoots back to two buds or leaves, and remove them back to their bases in summer. Tie the new shoots into the wire until they're each around 30cm (12in) from the central stem, then allow to grow upwards, tying into the cane or wire as they grow. Prune each stem as for a single cordon (see opposite).

For a multiple cordon, select three shoots in appropriate positions – one to each side and a central one. Train the central one as for a single cordon and the side shoots as for a double (above).

Redcurrants are especially attractive as multiple cordons.

SUMMER PRUNING

Summer pruning is carried out to concentrate the plant's energy into strengthening the main stem and producing good quality fruit, rather than lots of foliage. When the sideshoots are around eight leaves long, shorten them to five leaves long. Tie in the central stem(s) to the wires. Reducing the foliage allows more sun to get to the fruit to ripen it, and makes it easier for you to spot any problems that may arise with pests or diseases.

Shorten the sideshoots and tie in the central stem in summer.

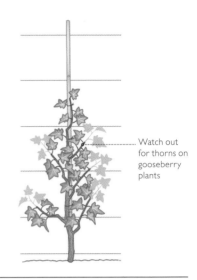

Watch out for thorns on gooseberry plants

WINTER PRUNING

Prune the new growth on the main, vertical stem back by a quarter. If the plant isn't putting on much new growth each year, cut the main stem back by half. This will stimulate more vigorous growth, but you'll still need to ensure the plant is adequately watered and fertilized (see p.26). Prune the main stem back to a bud on the opposite side to the previous winter's cut – this will keep the stem relatively straight over the years. Once the main stem reaches the top wire of the supports, prune back to one bud every year.

Sideshoots shortened in summer need pruning again to one or two buds from the main stem to prompt the production of more fruiting spurs. On older plants, some spurs bear less and less fruit each summer: prune these to make space for new, more productive spurs. Remove the Ds (see p.84) and any shoots growing towards the wall or fence; clear any shoots from the base of the stem.

Remove any shoots from the base of the stem when pruning in winter.

The buds for next year's growth are clearly visible on the bare stems.

If you plant some raspberry canes this autumn, you'll be picking fresh, sun-warmed berries for your breakfasts and desserts as soon as next summer.

CANES AND VINES

This section features cane fruits – plants such as raspberries and blackberries that fruit on tall, thin stems – as well as all the climbing plants, including kiwis, passion fruit, and grape vines. It also introduces the only annual fruits, melons and watermelons, which are sown from seed in spring and die after fruiting the subsequent autumn.

QUICK RETURNS

Annual and cane fruits are quicker than many plants to bear fruit after planting – either the same year (for melons, and some raspberries and blackberries), or the following year (for other raspberries, blackberries, and hybrid berries). The cane harvests won't be as heavy as in subsequent years, but they'll still be significant. The yields from mature plants are far larger than those you could buy for the same cost in the shops, and infinitely tastier for being homegrown and fully ripened in the sun. Passion fruit might not produce heavy harvests, but the pleasure of eating a homegrown fruit far outweighs its rarity.

VERTICAL HARVESTS

Cane fruits are naturally upright in their growth habit; climbers and grape vines need supports to grow successfully. These plants therefore all make the most use of the vertical space in your garden while taking up little horizontal space. Trained on a framework of freestanding posts and wires, they can form an edible hedge or screen, or utilize a bare wall or fence. They're fairly versatile in where they can grow, making them ideal for small gardens or narrow, awkward parts of the garden where there isn't space for much else, such as alongside pathways.

HOW BEST TO GROW

Cane fruits are fairly tolerant of a range of positions, although they'll fruit best in full sun and a rich, moist, but well-drained soil. Kiwis, melons, and passion fruit need more sheltered conditions, and passion fruit harvests are best when grown under cover. Grapes can be grown under cover, or outdoors if there are warm and fairly dry conditions from spring to autumn, and again fruit best in full sun. For the climbers and vines, you'll need a suitable structure onto which to train your plants – this could be a wall, fence, trellis, archway, or pergola. Alternatively, put in posts with taut wires running between them and train plants onto that. Posts and wires are also the best way to support cane fruits, although they can be tied into a wigwam of sturdy poles or wires fixed to a wall or fence.

KIWI *ACTINIDIA*

To grow kiwis of the size you'd buy in the shops, you'll need a large wall or fence on which to train them. However, dwarf plants, also known as cocktail kiwis, spread for only 2–3m (6ft 6in–10ft); they bear smaller fruits as well, no bigger than grapes, which can be eaten with the skin.

GROW FROM Potted plants
HEIGHT AND SPREAD Up to 5 × 5m (17 × 17ft)
HARDINESS Hardy
IDEAL SOIL Moist but well-drained
POSITION/SITE REQUIREMENTS ☼
TIME TO MATURITY Three to four years

CALENDAR

	SPRING	SUMMER	AUTUMN	WINTER
PLANT			▨ ▨ ▨	▨ ▨ ▨
HARVEST		▨	▨	

Through the year
Kiwis need a long season of warmth to fruit well, but they'll survive cold temperatures when dormant.

PLANT

Plant kiwis in autumn or winter (but not into frozen ground). Site them against a south- or southwest-facing wall or fence, or on a pergola or archway, putting in training wires or sturdy supports first. *Actinidia arguta* is hardier than *A. deliciosa*, but both types need a warm spot, sheltered from winds, to fruit well. Allow 4–5m (13–16ft) between *A. arguta* varieties, and 2.5m (8ft) between *A. deliciosa* types. If planting a male and a female, train them on the same structure, planting 60cm (24in) apart. Water well after planting, then mulch with compost or well-rotted organic matter, and cut back all shoots to 30cm (12in) tall.

IN CONTAINERS Kiwis aren't suitable for growing in containers, but they can be planted in a large raised bed at the foot of a wall and trained as described above.

A strong trellis is ideal for supporting vigorous kiwi plants.

> **TOP TIP** KIWIS ARE VIGOROUS PLANTS, BUT YOU CAN TURN THEIR RAMPANT GROWTH TO YOUR ADVANTAGE BY ALLOWING THE STEMS TO GROW OVER AN UNSIGHTLY SHED OR BIN STOREROOM.

GROW

Kiwis bear fruit on stems that are at least one year old, but their shoots become less productive as they age. Regular pruning both controls their size and promotes fresh growth. Kiwis can be trained as an espalier (see *pp.86–87*), but it's easier to tie in new growth to the wires regularly and prune out one or two four-year-old stems at the base each year during winter. Leafy side shoots can be shortened to three leaves from the main stem in summer. Check and loosen or replace old ties regularly and protect young shoots in spring with horticultural fleece.

Mulch in late winter with compost or well-rotted organic matter. Water young plants regularly, and mature plants in very dry spells.

Young kiwi shoots will twine themselves around wires or even their own stems.

VARIETIES

Kiwi plants are mostly either *Actinidia arguta* or *A. deliciosa*. There's little to tell between species or varieties in terms of taste, but *A. arguta* is a far hardier plant. Some are self-pollinating, others are either male or female, so you'll need one of each for good pollination and fruit on the female, but one male will pollinate up to eight females. Dwarf (cocktail) varieties mature more quickly than full-sized ones.

BRUNO (*Actinidia deliciosa*) Large, long fruits; sweet flavour; grow with 'Tomuri'.

HAYWARD (*A. deliciosa*) Reliable in temperate climates and a good flavour; grow with 'Tomuri'.

ISSAI (*A. arguta*) An early-ripening, self-pollinating, dwarf (cocktail) variety.

JENNY (*A. deliciosa*) Self-pollinating dwarf (cocktail) variety that can also act as a pollinator for other varieties.

JUMBO (*A. arguta*) Long, large, and sweet fruits; grow with 'Weiki'.

SOLISSIMO (*A. deliciosa*) Round, sweet fruits borne on hardy, self-pollinating plants.

TOMURI (*A. deliciosa*) Male plant for pollination.

WEIKI (*A. arguta*) Male plant for pollination.

Cocktail kiwis have skins that are less furry, so you can eat the entire fruit.

HARVEST

Pick kiwis when they're a little under-ripe – if left too long on the plant they'll become squishy and develop a slight taste of alcohol. Any fruit left on the plant should be picked before the first frost and ripened indoors.

Store kiwi fruits in a single layer in trays covered with a wax or plastic wrap. Kept at very cold temperatures (2–4°C/36–39°F), they'll stay fresh for four to six weeks. If ripe, they'll be slightly soft when gently squeezed.

Yields are variable depending on variety and the extent of late frosts.

It's easy to see and pick the fruit when kiwi plants are grown over a pergola.

MELON *CUCUMIS MELO*

Melons are annual crops, sown from seed and harvested in the same year, so your only outlay is a pack of seeds. Homegrown melons can be stored in the fridge for several weeks – but why do this when you have a delicious juicy fruit, still warm from the sun, in your hands?

GROW FROM Seed or young plants
HEIGHT AND SPREAD Up to 4 × 0.5m (13 × 1ft 9in)
HARDINESS Tender
IDEAL SOIL Moist but well-drained
POSITION/SITE REQUIREMENTS ☼
TIME TO MATURITY Three months

CALENDAR

	SPRING	SUMMER	AUTUMN	WINTER
PLANT				
HARVEST				

Through the year Melons need a warm spring and summer to fruit well, but are worth a try for the terrific rewards.

PLANT

Sow seeds in mid-spring, keeping the pots on a sunny windowsill or in a greenhouse. Plant out the seedlings in early summer once the roots are coming out the bottom of the pot and all risk of frost has passed. Ideally, grow them in a greenhouse or polytunnel, but these plants can be grown outside in warm temperate regions. They need a very sunny, warm, and sheltered site and a rich soil — mix in compost or well-rotted organic matter to the soil before planting. Space 1m (3ft) apart. Firm in plants to the soil well, and water after planting.

IN CONTAINERS Melons can be grown in medium-sized pots or growbags and allowed to trail, or they can be trained (see *opposite*) so they don't take up much space. Plant into a peat-free, multipurpose compost.

This seedling of 'Emir' has its seed leaves (the first pair to emerge) but not yet both true leaves.

ALSO TRY

Watermelons (*Citrullus lanatus*) can be grown in the same way as melons but, ideally, give them even warmer conditions. The variety 'Mini Love' has been bred for temperate climates and is fast to mature, with relatively small fruits. Young grafted plants are available – these are watermelon plants growing on the roots of another species, which is intended to transfer vigour and hardiness to the watermelon plant.

Growing in a nutrient-rich soil gives watermelons a fabulous flavour.

Seeds of heirloom varieties can be saved to sow next year.

Melons can be grown up the framework of a greenhouse.

GROW

Pinch out the top of the stem above the first two true leaves. You can then pinch out the tips of the two subsequent stems when they've reached seven leaves long, and the tips of the next shoots once they're five leaves long. The fruit will develop on the final set of shoots. Prune off excess leafy shoots throughout the summer.

Water regularly to keep the soil or compost consistently moist. Give plants in pots (and underperforming plants in the ground) a liquid feed fortnightly.

The stems can be either allowed to trail along the ground or trained up supports such as strings tied to the greenhouse roof, a strong cane or pole (either individual or as part of a wigwam), netting, or trellis. Tie in to the support regularly just below a leaf joint.

Thin the fruits to two to four per plant once they're around 2.5cm (1in) big. Support each melon on vertically grown plants to take the weight off the stem: pieces of netting or hessian are ideal; old bras or pairs of tights are allotment classics. Keep melons grown on the ground off the soil to prevent rot.

Pieces of hessian are ideal to construct a sling that will support the heavy, swelling fruits as they grow.

VARIETIES

There are three main types of melon: winter (including honeydew), cantaloupe, and musk, which differ in their flesh colour, size, and skin. Plant breeders have been developing new varieties specifically designed for cooler climates, with faster-maturing fruits on hardier plants.

ALVARO Aromatic cantaloupe melon that grows well outdoors. Award of Garden Merit (AGM, *see p.23*).

CHARENTAIS French heritage variety (*see p.23*) with sweet, orange flesh.

EMIR Cantaloupe melon that does well under cover in cooler, northern areas. AGM.

HONEY BUN Sweet, honey-fragranced melons are borne on compact and bushy growth.

OGEN Rich-flavoured flesh; fruits around 15cm (6in) in diameter. Grows best under cover; plants are compact. AGM.

HARVEST

Melons are ready to harvest when the skin begins to develop cracks. They'll sound slightly hollow when tapped gently and will also have a sweet aroma, although you might have to get up close to smell it. Take care not to break the stems when checking for ripeness. Each plant will yield two to four fruits.

Cut melons off the plant with sharp secateurs or a knife when ripe.

PASSION FRUIT *PASSIFLORA EDULIS*

It's worth growing passion fruit, also known as passion flower, for its gorgeous blooms alone, but the fruit is a magnificent bonus. This tropical species doesn't grow outside in temperate climates, but it does make a beautiful houseplant or an attractive specimen for a greenhouse or conservatory.

GROW FROM Potted plants
HEIGHT AND SPREAD Up to 8 × 4m (26 ×13ft)
HARDINESS Tender
IDEAL SOIL Moist but well-drained
POSITION/SITE REQUIREMENTS ☼
TIME TO MATURITY Two years

CALENDAR

	SPRING	SUMMER	AUTUMN	WINTER
PLANT	▓			
HARVEST		▓	▓	

Through the year Passion fruit are evergreen, and flower from early summer to autumn.

PLANT

Passion fruit are climbers that fruit on the sideshoots that grow off the plant's main stem. They need training on wires fixed to an interior or greenhouse wall, or some other indoor framework. It's easiest to train two main stems up and along the top of a wall (or other support), then allow sideshoots to develop and hang downwards, covering the wall. You can also train them over windows and doorways to create a living curtain of foliage and flowers.

The plants need full sun and minimum temperatures of 10°C (50°F) in winter and at least 20°C (68°F) in summer to fruit well. They also need a quite humid atmosphere. Yellow-skinned passion fruit prefer even warmer temperatures. Allow each plant 3m (10ft) of space, planting into containers (*see below*).

IN CONTAINERS Use a container at least 35cm (14in) in diameter, filled with a peat-free, soil-based compost mixed with grit in a ratio of 3:1. Place pots in a saucer or tray to catch any water running out from the bottom.

Passion fruit thrive when planted if, as here, they have healthy, well-developed root systems but aren't rootbound.

The stems of passion fruit can be fixed to a wire attached to the top of a wall and allowed to trail downwards.

TOP TIP TO GROW NEW PASSION FRUIT PLANTS FROM SEED, PUT SOME SEEDS AND FRUIT PULP IN A SAUCER AND SET ASIDE FOR A FEW DAYS. THEN WASH, DRY, AND SOW THE SEEDS JUST BELOW THE SURFACE IN SMALL POTS OF PEAT-FREE COMPOST, KEEPING THEM IN A VERY WARM PLACE FOR THE BEST RESULTS.

Passion-fruit flowers need hand-pollinating to ensure good fruit yields.

Passion fruit develop over a long period, so check the plant regularly for ripe fruit that's ready to pick.

GROW

Water passion fruit plants regularly, but ensure the compost doesn't become sodden in winter to avoid root rot. Add a liquid feed to the water every four weeks during spring and summer. If the atmosphere is relatively dry, mist the leaves with a spray bottle of water every day to help keep the humidity high around the plant. Add a thin layer of mulch (fresh compost) to the top of the container in late winter, making sure it doesn't touch the stems.

Flowers will need hand-pollinating with a soft paintbrush: gently brush each flower on a sunny afternoon to spread the pollen between the flowers. Repeat this regularly to ensure all new flowers get pollinated. In very hot weather, the plants will benefit from being shaded from the worst of the afternoon sun as their leaves can easily become scorched.

Prune the tips of pendulous sideshoots as necessary to keep them off the ground; in spring, prune old fruited stems back to 20cm (8in) long. Plants become tired relatively quickly, and will need replacing every five or six years.

VARIETIES

Passiflora edulis is the most reliable passion fruit and has purple-skinned fruit; *P. edulis* var. *flavicarpa* has yellow-skinned fruit. There are some named varieties of passion fruit, but very few are widely available – the species itself should be sourced from a specialist fruit nursery – and there's not much difference between them.

Yellow passion fruit can be twice as large as the purple-skinned alternative.

HARVEST

Ripe fruits are fully coloured and slightly wrinkled. Pick them by cutting the stalk with sharp scissors or secateurs. They can be ripened a little further in a warm room or stored for a week or more in a container in the fridge. Yields are variable – grown indoors, passion fruit provide an occasional delicacy rather than bumper crops.

The more wrinkled the skin of the passion fruit, the drier and sweeter the pulp inside will be.

BLACKBERRY *RUBUS FRUTICOSUS*

Although foraging for wild blackberries is fun, growing your own means you can grab a handful whenever you want them from fruit that's easy to reach on neatly trained canes. The cultivated varieties have other advantages too: the berries are generally larger, and some canes have no thorns at all.

GROW FROM Potted or bare-root canes
HEIGHT AND SPREAD Up to 2.5 × 4m (8 × 13ft)
HARDINESS Hardy
IDEAL SOIL Moist but well-drained
POSITION/SITE REQUIREMENTS ☼ ◑
TIME TO MATURITY Two years

CALENDAR

	SPRING	SUMMER	AUTUMN	WINTER
PLANT				
HARVEST				

Through the year
Blackberries' pretty flowers are popular with bees and other pollinating insects.

If you have space, blackberries can be left to ramble over low fences.

PLANT

Plant in autumn, ideally, or winter if the ground isn't frozen. Space plants at least 2.5m (8ft) apart, up to 4.5m (15ft) for vigorous varieties. Fix the training wires (and posts) or trellis in place first (see *p.136*), or plant where they can be allowed to ramble over an archway, low fence, or similar. Blackberries fruit best in full sun but tolerate dappled or partial shade; in dry and hot areas they prefer some shade. Water well and apply a mulch of compost or well-rotted organic matter after planting, then cut back all the canes to ground level to stimulate production of new shoots.

IN CONTAINERS Only the dwarf varieties specifically bred for pots are suitable for growing in containers. Plant into soil-based, peat-free compost, repotting each spring into a bigger container until the pot is about 50cm (20in) in diameter.

GROW

Tie in the canes to the wires as they grow, spacing them so they all get as much sun as possible (see *p.137*). Mulch with compost or organic matter mixed with a little wood ash in late spring and water young plants regularly; mature plants will need watering in dry spells. Blackberries mostly fruit on canes produced the previous year: prune in late autumn or early winter (see *p.137*) and tie in the new shoots to protect them from winter winds. Some ("primocanes", or first canes) fruit on the current year's growth; prune as for autumn raspberries (see *p.136*). Plants can be affected by aphids, and birds will try to eat the ripening fruit (see *p.38*).

Wear gloves to avoid thorns when tying in your blackberry canes.

New Zealand 'Karaka Black' has an excellent, floral flavour.

VARIETIES

Garden varieties of blackberries tend to be sweeter than the wild fruit. By planting a few varieties, you can extend the season from midsummer well into early autumn. All are self-pollinating.

APACHE Early season. Thornless, upright growth and large berries.

ASHTON CROSS Mid- to late season. Excellent for jam; fruits taste similar to wild blackberries.

BEDFORD GIANT Early season. Vigorous plants with large, soft, tasty fruits.

KARAKA BLACK Early to late season. Enormous, elongated fruits ripen over a long period.

LOCH NESS Mid-season. Upright canes and compact habit, favoured by commercial growers. Award of Garden Merit (AGM, see p.23).

LOWBERRY LITTLE BLACK PRINCE Mid- to late season. Extended harvests of sweet fruits are borne on this dwarf variety bred for containers.

OREGON THORNLESS Mid- to late season. Canes are thorn-free; well-flavoured fruits and good autumn colour.

PURPLE OPAL Early to mid-season. Dwarf variety for growing in pots; thornless.

REUBEN Early season. A primocane (see *opposite*) with compact, upright growth, and flavourful fruits.

SILVAN (syn. 'Sylvan') Early season. Plants are tolerant of heavy or poor soils and dry conditions.

Ripe blackberries will be ready to pick every few days.

HARVEST

Blackberries are ready to pick when the entire berry is a rich purple-black colour and will detach easily from the cane. The fruits ripen gradually over a period of a few weeks or even months. Ripe berries will keep for a short time in the fridge, if necessary. Alternatively, they're great for either freezing (see p.46) or cooking.

Expect to be able to pick 4–6kg (9–13lb) or more per mature plant, depending on the vigour of the variety; dwarf potted varieties yield significantly less fruit.

RASPBERRY *RUBUS IDAEUS*

Research suggests that raspberries grown outdoors are far tastier and more nutritious than those grown in polytunnels, like most commercially produced berries. They thrive in deep, rich soils, but even in a small space you can have fresh berries by growing dwarf varieties suitable for containers.

GROW FROM Potted or bare-root canes
HEIGHT AND SPREAD Up to 2.5 × 0.6m (8 × 2ft)
HARDINESS Hardy
IDEAL SOIL Moist but well-drained
POSITION/SITE REQUIREMENTS ☼ ☼
TIME TO MATURITY 6–18 months

CALENDAR

	SPRING		SUMMER		AUTUMN		WINTER	
PLANT	░				░	░	░	░
HARVEST			░	░	░	░		

Through the year
By planting several different varieties, you could be picking fresh raspberries all summer and well into autumn.

PLANT

Ideally, plant bare-root or potted raspberries during the autumn, but you can also plant them in the winter or early spring, as long as the ground isn't frozen. Be sure to choose a sheltered spot to avoid wind-damage to the canes. Raspberries will tolerate dappled or partial shade, but they'll fruit better in full sunlight. Spread a 5cm (2in) thick layer of compost or well-rotted organic matter over the planting area and dig it in lightly before planting. After planting, add another layer as a mulch.

Summer-fruiting raspberry canes will need the support of wires or posts (see pp.136–137); this is beneficial but not absolutely essential for autumn-fruiting varieties. In a row, space the plants 45cm (18in) apart.

IN CONTAINERS Smaller and dwarf/patio varieties of raspberries will do well in containers. Pot-growing is also a good option if you have alkaline or chalky soil (see p.15), in which raspberries won't thrive. Use a 1:1 mix of multipurpose or garden compost and a soil-based compost (both peat-free) and a large container, around 50cm (20in) in diameter.

Use long, rectangular containers to plant a row of raspberries that will act as an effective screen.

Plant bare-root canes as soon as possible after buying them.

You can tie raspberries onto a fence for support, which saves space in the garden.

GROW

Water newly planted canes and those in pots regularly, and mature plants in dry spells. Give plants in pots a fortnightly liquid feed during spring and summer. Mulch annually with compost or well-rotted organic matter. Prune in summer or late winter, depending on variety (see right and pp.136–137), and tie in canes to the support, as necessary. Raspberries will "sucker", growing under the ground and putting up new shoots: thin out the weakest and those growing where you don't want them when pruning to leave a row of canes about 10cm (4in) apart.

HARVEST

Raspberries are ripe when they're a rich red (or gold) colour and pull easily off their central plug – the ripest fruits will fall at the merest nudge of the stem so pick them regularly to avoid half the harvest dropping to the ground.

Summer-fruiting varieties can yield around 2.5kg (5lb 8oz) of fruit per metre (3ft) of row, autumn-fruiting varieties around 1kg (2lb).

Raspberries are at their ripest and sweetest when they're richly coloured and warmed by the sun.

VARIETIES

Summer-fruiting raspberries develop fruit on canes produced the previous year (floricanes); autumn-fruiting varieties fruit on canes that are produced earlier in the same year (primocanes). Primocanes are more compact and straightforward to prune than floricanes, but have lower yields – although by planting both types, you can extend the harvest until the first frost.

ALL GOLD Primocane. Late season variety. Yellow berries have excellent flavour; a compact plant. Award of Garden Merit (AGM, see p.23).

AUTUMN BLISS Primocane. Late season. The best autumn-fruiting raspberry available, producing good yields of tasty berries. AGM.

GLEN AMPLE Floricane. Mid-season. Produces heavy yields of good-flavoured fruit. AGM.

GLEN COE Floricane. Mid-season. Aromatic, black-purple fruits produced on spine-free canes with minimal suckering

GLEN FYNE Floricane. Mid-season. Compact plants produce excellent aromatic, firm berries.

JOAN J Floricane. Mid- to late season. Heavy yields of berries over a long period; a traditional raspberry taste. AGM.

MALLING JEWEL Floricane. Early season. Compact growth and good yields of large, firm berries. AGM.

RUBY BEAUTY Floricane. Mid-to late season. Dwarf plant ideal for containers; good yields of sweet berries.

TULAMEEN Floricane. Mid- to late season. Vigorous plants produce large, tasty berries over a long period. AGM.

Raspberry 'All Gold' has a sweet raspberry flavour and an unusual colour.

HYBRID BERRIES *RUBUS* SPECIES

Hybrid berries (such as loganberries and tayberries) are fruits produced from cross-breeding raspberries, blackberries, and each other. They have a great range of flavours, sizes, and colours and are easy to grow; care and cultivation is similar to blackberries, but many are less vigorous and thornless.

GROW FROM Potted or bare-root canes
HEIGHT AND SPREAD Up to 2.5 × 0.6m (8 ×2ft)
HARDINESS Hardy
IDEAL SOIL Moist but well-drained
POSITION/SITE REQUIREMENTS ☼ ☀
TIME TO MATURITY 6–18 months

CALENDAR

	SPRING	SUMMER	AUTUMN	WINTER
PLANT				
HARVEST				

Through the year
Hybrid berries are mostly early summer crops, but the harvest is spread over several weeks.

Tayberry plants can be tied into wires on a fence or the side of a shed.

PLANT

Ideally plant new canes in autumn, or in winter if the ground isn't frozen. Space plants around 2.5m (8ft) apart. They can be allowed to ramble over a low fence or archway, or trained onto a vertical surface with wires (and posts) or trellis (see p.136). Hybrid berries will tolerate dappled or partial shade but fruit best in full sun; they prefer some shade in dry and hot areas. Water well and apply a mulch of compost or well-rotted organic matter after planting, then cut back all the canes to ground level to stimulate the production of new shoots.

IN CONTAINERS Hybrid berries aren't suitable for growing in containers, although it's possible that in future plant breeding will introduce dwarf varieties that can to be grown in pots.

GROW

Water young plants regularly; mature plants need watering in dry spells. If you're training the plants, tie in the canes to wires as they grow, spacing them so that they all get as much sun as possible (see p.137). Hybrid berries bear fruit on canes produced the previous year: prune in late autumn or early winter (see p.137) and tie in the new shoots to protect them from breaking in winter winds. Mulch with compost or organic matter mixed with a little wood ash in late spring.

Check black raspberries (see right) regularly for newly ripened fruits.

VARIETIES

Some hybrid berries are just a single species, others have a small choice of varieties within the species. All are self-pollinating; all are early summer crops, except loganberries, which ripen over a long period in the summer.

BLACK RASPBERRY JEWEL (*Rubus occidentalis* 'Jewel') Productive canes are an ornamental white; the delicious fruit is black in colour and high in antioxidants.

BOYSENBERRY (*R.* 'Boysenberry') A heritage berry (see *p.23*), the result of complex cross-breeding of the berry family. Large, tasty, purple-black berries are good both cooked and fresh.

DEWBERRY (*R. caesius, R. ursinus*, and other *Rubus* species) Small black fruit with a white bloom.

LOGANBERRY LY 59 (*R.* × *loganobaccus* 'Ly 59') Long red fruits taste like an acidic cross of strawberries and raspberries; excellent for jam. Ly 654 (*R.* × *loganobaccus* 'Ly 654') is less vigorous than 'Ly 59' and thornless; Award of Garden Merit (AGM, see *p.23*).

TAYBERRY (*R.* Tayberry Group) Long fruits are sharp, with a fruity flavour good for cooking; AGM. 'Buckingham' (*R.* (Tayberry Group) 'Buckingham') is a thornless variety.

TUMMELBERRY (*R.* 'Tummelberry') A red berry that's best used for jam but can also be eaten fresh. Plants tolerate colder climates well.

Loganberries can be eaten fresh but they're also delicious cooked.

Bright red tummelberries are ready to pick in early summer.

Dewberries are the smallest of all the hybrid berries.

Boysenberries have an intensely fruity, wine-like flavour.

ALSO TRY

Japanese wineberries (*Rubus phoenicolasius*) were introduced into Europe in the 1890s. Their attractive canes are thickly covered in shiny red spines, but these are soft and furry rather than painful to touch. The berries themselves are borne on the ends of the stems (which makes them easy for picking) and are scarlet red, juicy, glistening, and with an excellent grapey-raspberry flavour.

Japanese wineberries are also known as Chinese blackberries.

HARVEST

Hybrid berries are ready to pick when they have fully developed their colour and detach easily from the cane. They'll keep for a short period in the fridge if necessary. Alternatively, freeze or cook them (see *p.46*). Expect to pick 4kg (9lb) or more per mature plant depending on the vigour of the variety. Check the fruits and wash them well as hybrid berries are vulnerable to raspberry beetle (see *p.39*). Remove and dispose of any affected berries when picking to prevent re-infestation.

GRAPE *VITIS VINIFERA*

Whether you'd like a backyard vineyard or just to grow a few grapes for your cheeseboard, grape plants will fit almost anywhere. Their gnarled framework can be kept quite compact when trained onto a wall or allowed to scramble in a more unruly fashion over a pergola.

GROW FROM Potted plants
HEIGHT AND SPREAD Up to 6 × 1m (20 × 3ft)
HARDINESS Hardy
IDEAL SOIL Moist but well-drained (also grows well in drier soils)
POSITION/SITE REQUIREMENTS ☼
TIME TO MATURITY Five years

CALENDAR

	SPRING			SUMMER			AUTUMN			WINTER		
PLANT		▓	▓							▓	▓	▓
HARVEST							▓	▓				

Through the year Grape vines are deciduous, and have beautiful autumn colour in shades of yellow and red.

PLANT

Ensure your plants are grafted onto a *V. labrusca* rootstock and are certified as free from a pest called phylloxera.

Grape vines need a sheltered spot in full sun, but with some ventilation to help prevent mildew (see p.41). They won't thrive on chalky soil. Ideally, plant in late spring, after all risk of frost has passed, watering well and mulching with compost or well-rotted organic matter. If you want to grow your vines inside a greenhouse, planting them outside and training them through a hole in the frame is the best option, but they can be planted in a container (see below).

Vineyards grow their vines according to a system of pruning called Guyot, and there are other traditional techniques as well. However, it's just as productive, and far simpler, to grow grapes as single or multiple cordons. Cordons can be used for growing grapes outdoors and under cover, and gives you a more architectural winter structure to look at as well. Put in supporting posts, spaced 4m (13ft) apart, holding at least three horizontal wires, 30cm (12in) apart (see also pp.136–137). You can also grow them up a wall, fence, or the interior framework of a conservatory or greenhouse: fix the wires as far off the surface as possible, spaced vertically, as above. When planting more than one vine, allow 1.2m (4ft) between the plants.

IN CONTAINERS Grapes grown in a greenhouse can be planted in containers. Plant your vines into a container of peat-free, soil-based compost at least 50cm (20in) in diameter. The container will need to be placed at the base of a vertical surface against which the vine can be trained and supported.

Grape vines can be grown as high as you like against a sunny wall – just make sure you can reach to pick them.

Grapes grown inside a greenhouse can be planted in containers.

GROW

Water young plants as they establish for the first year or two, and thereafter only in dry spells. Give plants an annual mulch of compost or well-rotted organic matter. Vines in containers will need regular watering and a fortnightly application of a liquid fertilizer.

As your vine establishes, it's best not to let it divert its energy into producing fruit. A sacrifice now will result in a healthier, stronger plant that will produce bigger yields later. Remove all flowers for the first two years after planting; allow three bunches on the plant for the following two years; it can then be allowed to flower and fruit normally. However, removing all but one bunch per 45cm (18in) of vine will give you bigger, tastier grapes.

On mature vines, as the grapes develop, take off any leaves shading the bunches of grapes as this will speed up ripening, reduce pest and disease incidence, and boost the flavour and nutrient content of the grapes. Mildew and grey rot can infect plants, and wasps enjoy the fruit. Look out for yellowing leaves with green veins in summer – this is a sign of magnesium deficiency and can be remedied with the application of a balanced liquid fertilizer. For pruning and training information, *see pp.140–141.*

Cut the bunch of grapes along with a short section of stem, which you can use as a handle.

HARVEST

Grapes don't ripen further once picked, and also don't ripen evenly across the bunch (the top, or shoulders, of the bunch will ripen first), so it's always best to wait as long as you can before harvesting them. The majority of the grapes on the bunch will be fully coloured and slightly soft when they're ripe, and the green stalk will have withered slightly to a straw-yellow colour. Pick by cutting so that some of the stem is still attached to the bunch. Freshly picked grapes will keep for around two weeks when stored in the fridge. Yields are variable and depend on the size of the vine.

The leaves – for dolmades and other culinary uses – can be picked in spring: those about three leaves down from the tips of young shoots are the best.

Cutting back the foliage growth helps the fruit to ripen.

'**Boskoop Glory' grapes** contrast with the striking colour of the autumn leaves.

'**Muscat of Alexandria**' is one of the oldest grape varieties.

Colourful roses will help to attract pollinators to your vines.

> **TOP TIP** COPY THE VINEYARDS AND PLANT ROSES NEAR YOUR VINES IF YOU'RE GROWING A LOT OF THEM (*SEE ABOVE*). THE ROSES ARE SUSCEPTIBLE TO MANY OF THE SAME DISEASES AND PESTS AS THE VINES AND TEND TO GET AFFECTED FIRST, GIVING YOU ADVANCE WARNING TO TAKE PREVENTATIVE MEASURES.

VARIETIES

There are two main choices to make when selecting a grape variety to grow: do you want wine or dessert grapes, and do you want white (green) or black (red) grapes? Wine grapes have higher tannin levels (giving them more flavour but also bitterness and astringency), thicker skins, and seeds. Dessert (eating, or table) grapes are generally seedless, thin-skinned, and larger, but more watery. Wine grapes are good to eat fresh too – they tend to be sweeter than dessert grapes – and the plants themselves are far better suited to growing outdoors. Only wine grapes are suited to making wine; but note that it's illegal to sell alcohol without a licence. Black grapes are higher in healthy antioxidants and phytonutrients.

DESSERT GRAPES

BOSKOOP GLORY Black. Late season. A reliable cropper outside with good flavour. Award of Garden Merit (AGM, see p.23).

BRANT Late season. Small, tight bunches of grapes; plants have some mildew resistance and good autumn colour.

CHASSELAS (syn. 'Chasselas d'Or') White. Early season. Best grown under cover, but can also be grown outdoors; pale golden green, juicy grapes.

FOSTER'S SEEDLING White. Early season. Needs the protection of a greenhouse; produces large bunches of amber grapes.

HIMROD White. Late season. Seedless grape, with a honeyed flavour; productive plants need the protection of a warm and sunny wall.

MUSCAT OF ALEXANDRIA White. Late season. Excellent flavour but best grown under cover.

NEW YORK MUSCAT White. Late season. Excellent flavour but needs a sheltered and warm spot to fruit well. AGM.

SCHIAVA GROSSA (syn. 'Black Hamburgh') Black. Mid-season. Fruits reliably under cover; good crops of grapes with an excellent flavour.

WINE GRAPES

BACCHUS White. Late season. Aromatic grapes also suit roasting with savoury foods and making grape jelly.

CABERNET CORTIS Black. Late season. A herby, blackcurrant flavour, rich in tannins; plants have some disease resistance.

FRAGOLA (syn. 'Isabella' and 'Strawberry Vine') White. Mid- to late season. Musky, strawberry-like aroma; grapes have a dark red or purple hue.

GEWÜRZTRAMINER White. Late season. Excellent flavour and red-pink colour; grows well in cool climates but prone to frost damage, pests, and diseases.

LEON MILLOT White. Late season. Excellent flavour and red-pink colour; grows well in cool climates but prone to frost damage, pests, and diseases.

MADELEINE ANGEVINE White. Mid- to late season. A later-flowering and heavy-cropping vine that's suited to cooler areas.

MÜLLER-THURGAU White. Late season. Reliable harvests of sweet, juicy grapes; bred from the Riesling grape.

PINOT NOIR Black. Mid- to late season. Heavy yields of juicy grapes even on poor soils; excellent flavour.

Mature 'Müller-Thurgau' vines will fruit prolifically every year.

Pink 'Gewürztraminer' grapes have a delicious aroma of roses and spice.

PRUNING CANE FRUITS

Pruning cane fruit is easy – it simply involves cutting off the stems at ground level. However, you need to know when to prune the two main types of fruit to avoid losing your crop. The first types of fruits are those that are produced in summer on canes that grew during the previous summer or autumn. These are known as floricanes, and include summer-fruiting raspberries, blackberries, and hybrid berries. The other type – primocanes – fruit in autumn on canes produced in the same year. These include autumn-fruiting raspberries.

Support blackberries, such as this cut-leaved variety, on fences of any height by bending the canes to grow horizontally.

SUPPORTING CANES

Floricanes will always need tying in to supports to prevent their stems from breaking in winter winds and to keep their stems properly under control. The canes can either be tied into wires that are fixed to a wall or fence (see pp.86–87), or grown in rows supported and kept neat by a system of posts and wires. These structures may have single wires or double wires, which require slightly more elaborate supports (see right).

Whether you're using single or double wire supports you'll need a sturdy post at least every 10m (33ft), and to support the posts with diagonal buttress posts. Ensure your posts are sunk to a depth of at least 60cm (24in).

Most primocanes don't need supports but they may benefit from being staked to bamboo canes to maintain a tidy appearance.

> **TOP TIP** USE A FIGURE OF EIGHT KNOT TO TIE IN CANES, AS IT'S SECURE BUT WON'T CHOKE THE CANES AS THEY THICKEN AND GROW (SEE P.31).

Creating double wire supports for your canes is a bigger initial investment, but less work in the long run.

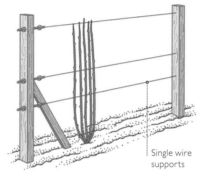

Single wire supports

Single wire supports: add wires at 75cm (30in), 1.2m (4ft), and 1.5m (5ft) from the ground.

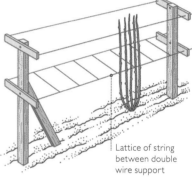

Double wire supports: install lower wires at 1m (3ft 3in) and top wires at 1.5m (5ft) from the ground.

Lattice of string between double wire support

TYING IN NEW GROWTH

If you opt for a single wire support system, and if you are growing against a wall or fence, you will need to tie new canes individually to the wires as they grow, spacing them apart to allow light, air, and pollinators to reach each cane. Keep the new growth of floricanes separate from the fruiting canes as you tie them in over the summer, this will make life easier when it comes to harvesting and pruning. With wall-trained fruit – especially the long canes of blackberries – you can try more creative arrangements, such as tying the new canes in a fan shape.

If you've built a double wire support, you need only to guide canes between the wires. You can tie a lattice of string between the two wires (see *left*), creating "boxes" to space the canes more evenly, but it's not essential.

After pruning out the old canes in autumn, the new growth needs a final tying in to stop it breaking in winter winds. If the canes are long enough, the tops can be bent over and tied down in a loop, which will stimulate more buds to break, giving a bigger harvest next year.

Bending over and tying in the tops of the canes looks good and encourages more fruit buds.

PRUNING PRIMOCANES

Simply cut all the canes of primocane varieties back to the ground during early winter after they have finished fruiting, leaving no stubs that could become infected by rot. You can sometimes induce a double crop from autumn raspberries by leaving all the canes unpruned after their first summer's crop. The old canes should crop in the next summer.

Prune all the canes of autumn-fruiting raspberries and hybrids to the ground.

PRUNING FLORICANES

Prune floricanes in autumn, after the last harvest. Cut the old, brown canes (from which you picked in summer), leaving the new green canes to bear fruit next year. Cut any spindly or broken new growth as well; aim for six to eight canes of new growth per plant. Just cut the canes back to the ground, leaving no stub to harbour rot.

It's easy to tell the difference between the brown old canes and the green new growth on floricanes.

GROWING CLIMBERS

Climbers such as passion fruit and kiwi can be used to disguise an ugly fence, create living screens, or shade a sunny spot indoors (passion fruit) or out (kiwi) – it's a great way to add more plants to the garden, and they'll give you a harvest too. Utilizing the vertical space also creates wonderful homes and shelter for wildlife, and climbers on walls can help insulate your house.

The twisty stems of kiwi plants will entwine themselves around the support.

SUPPORTS FOR CLIMBERS

There are a number of ways climbing plants can be supported, including by a pergola or trellis, a system of wires fixed to a wall (see pp.86–87), or wires on the interior frame of a greenhouse or conservatory. However you choose to grow your plants, it's important to remember that the plants can get very heavy (especially kiwi vines), so the supports must be sturdy and strong enough to take the weight. Don't let them grow out of reach so that it's impossible to access the fruit at harvest time. A few minutes of your time once or twice a week to tie in new shoots if necessary, and a good annual pruning, keeps them from getting out of control (see pp.120–135).

A pergola is an ideal support for growing kiwi vines.

Water new fruit plants especially well when they're planted at the bottom of walls or fences.

PLANTING CLIMBERS

The procedure for planting climbers is much the same as for any other plant, but with a few caveats (see pp.30–31). Put up support systems before planting, then plant 15–20cm (6–8in) away from the support. It helps the climber to get up onto the support if you angle the plant slightly in the planting hole, to point it towards the support. If it's not yet quite tall enough to reach,

Young fruit plants can be grown up a cane in a container until your supports are ready.

use a cane or some string to guide the growth upwards, removing it (and any canes that came with the plant) once it has got a hold on the support. The second thing to be aware of is that the soil at the base of walls and fences is often significantly drier than the rest of the garden because the wall creates a "rain shadow", an area where the rain can't reach. Plants trained onto walls and fences will therefore need more watering than those in the open.

Maximize the vertical space at the back of a border by training a kiwi up the fence.

PRUNING GRAPES

Grape vines grown for wine are typically pruned and trained according to the tricky Guyot system, but growing vines as cordons (known as the "rod-and-spur" system, described here) is far simpler, and can be used to create productive vines outdoors and under cover. If you'd prefer to leave your vines to scramble of their own accord over a pergola, just prune back the excess growth when it gets in the way.

Grown as cordons, the plants can be trained to suit your needs, as with this raised screening.

CORDON: YEAR 1

Vines planted in late autumn or early winter should be pruned to a strong bud at a height of about 30cm (12in) high, level with the bottommost wire of the support. Vines planted any later should be left unpruned because they'll weep sap if cut. In the summer, tie in the central shoot as it grows. Cut back sideshoots to five or six leaves long, and any shoots off those to just one leaf long.

A neatly trained cordon encourages good air circulation around the vine, and so reduces the likelihood of disease.

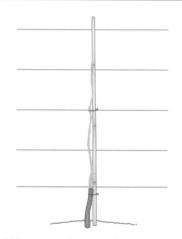

Winter pruning, year 1: cut back to the first wire.

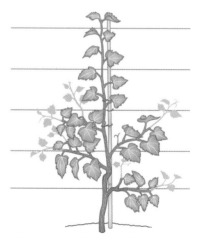

Summer pruning, year 1: cut back sideshoots to six leaves.

CORDON: YEAR 2

In late autumn of the second year, cut back the central shoot by around half of its new growth, removing all the green part of the shoot and leaving only the brown, hardened (ripened) woody part. Cut back all sideshoots to one or two buds. In summer, prune and tie in as for summer in year one and remove all the flowers (see p.133). In subsequent years, continue to cut the central stem back by half of the new growth each winter until the ripened stem is as tall as the topmost wire. Each time, cut to a bud on the opposite side of the stem from the previous year's cut, to keep growth relatively vertical.

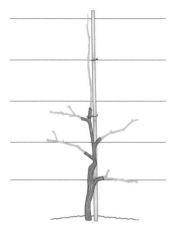

Winter pruning, year 2: cut back leader as soon as the leaves fall.

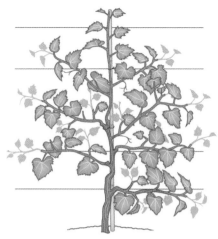

Summer pruning, year 2: cut back laterals to six leaves. Do not let fruit form.

A well-pruned established vine starting back into growth in spring.

PRUNING AN ESTABLISHED CORDON

In winter, shorten all the sideshoots to two buds and cut out the sideshoot stubs (spurs) where they have become congested. Once the sideshoots start to grow, remove all but two from each spur – one to fruit and one as a back-up. Once they're growing, select the best to tie in for that year's fruiting growth and cut the back-up down to two leaves.

In summer, shorten non-flowering sideshoots to five leaves and cut flowering sideshoots to two leaves beyond the last flowers on the shoot. Shoots off the sideshoots should be shortened to one leaf. Tie in all sideshoots and the stem; remove excess flowers (see pp.132–133).

To encourage new buds to break from an older vine, prune the central stem to a bud just below the top wire. Untie it from the support in early spring and carefully bend to a horizontal position (or as close as you can get), then tie it down. Once new buds are breaking, untie and secure the stem back in its vertical position.

In spring, remove all but two shoots from each spur.

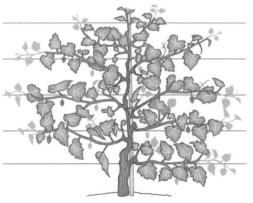

In summer, tie in the leader and shorten laterals lacking flower trusses to six leaves.

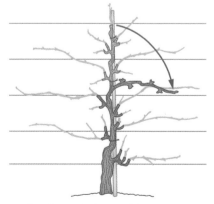

In winter, tie the top of the leader so that it's horizontal. Release it in spring.

INDEX

Bold text indicates a main entry for the subject.

Author Holly Farrell

AUTHOR ACKNOWLEDGMENTS

Many thanks to Diana Loxley, Paul Reid and Marek Walisiewicz – it was great working with you all, and a lot of fun – and to Chris Young for his suggestion in the first instance.

PUBLISHER ACKNOWLEDGMENTS

DK would like to thank Mary-Clare Jerram for developing the original concept, Chris Young for content origination, Nicola Powling for jacket development, Margaret McCormack for indexing, and Paul Reid, Marek Walisiewicz, and the Cobalt team for their hard work in putting this book together.

PICTURE CREDITS

The publisher would like to thank the following for their kind permission to reproduce their photographs:

Alamy Stock Photo: A.D.Fletcher 72bl; ableimages 63bl; agefotostock 114tr, 122br; Alexander Chizhov 4c; Alexandru Adrian Ionas-Salagean 11bl; All Canada Photos 135bc; Amelia Martin 93tr; Amelia Mortimer Photography 44cr; Anja Schaefer 81cl; Anna Sedneva 4c; Anna Usova 32cl; Arterra Picture Library 42bl; Avalon.red 135tr; Bax Walker 131tc; beatrice preve 11tr; Bellephoto 95br; Besjunior 48tr; BIOSPHOTO 97bc, 136tr, 138cr; blickwinkel 16br, 69tl, 69br, 71tc, 74tl, 81br, 129tl, 140tr; Bojan Živković 41tl; Botany vision 111br; Carl Newman 35tc; Cavan Images 8tr; christopher miles 16bl, 76bl; Clare Gainey 12tr; Cora Müller 55br; Dave Bevan 38tr, 41br, 130cl; Deborah Vernon 6c, 21cl, 23bl, 27t, 31tl, 46bl, 65cra, 82tr, 84cl, 90c; dinh ngoc hung 124cr; dpa picture alliance 53tl; Ed Callaert 56cl; Edd Westmacott 65bl; Erkki Makkonen 25c; EyeEm 73tl,138tr; Felix Choo 67br; Fir Mamat 28tr; flowerphotos 131tr; Garey Lennox 125bc; George Robinson 47c; Gheorghe Mindu 92cr, 124c; GKSFlorapics 42tr, 111tl; H-AB 125tr; HD57flora 39bl; Heiko Küverling 132bl; Holmes Garden Photos 134cl; Igor Normann 71br; Imagebroker 126cl; Ingrid Balabanova 133bl; Island Images 32br, 46bc; Jacquelin Grant 130br; Janet Horton 8cr,136br; Jason Smalley Photography 33c; Jim Allan / Stockimo 79tl; Joe 74cr; John Keates 75br; JOMWASCHARA KOMVORN 125cr; Juniors Bildarchiv GmbH 12br; Jürgen Kottmann 43tc; Karen Kaspar 9bl; Kate Branston 134tr; Kay Roxby 99bc; Klaus Oskar Bromberg 21tl; Koba Samurkasov 57bl; Larisa Blinova 14cl; Lea Rae 37tl; Lesley Pardoe 16tr; LianeM 97tl; 138c; Magelanic Clouds 138bl; Maggie Sully 68tr; Mark Bolton Photography 84tr; Martin Hughes-Jones 80br; Matthew Taylor 71tl, 101tr; mauritius images GmbH 28br, 44tr, 44bl; MBI 26tr; McPhoto/Rolf Mueller 105bc; mediasculp 48cl; mike jarman 70bl, 107tr; Mitchel Hutchinson 15tr; Moggara12 56cr; Naturefolio 78cr; nevenm 30br; Nigel Cattlin 39tc,40tr, 40br; Oleksandr Berezko 128br; Oleksandr Bushko 127tl; Oleksandr Rado 28cl; Oleksii Terpugov 122cr; P Tomlins 134tr; Panther Media GmbH 57tr, 120bc; Peter Turner 140bl; Phanie 98cr; Premium Stock Photography GmbH 131cl; PURPLE MARBLES GARDEN 22cl; Reppans Horticulture 117br; rjp 113br; RM Floral 81tl; ShawStock 88tr; Stephen Dorey 60bl; Steve Taylor ARPS 88bl; Steven May 9tr; STUDIO75 112bl; Thanh Thu Thai 23tc; Tim Gainey 20tr, 74tr, 95tr,123tr;Tomasz Klejdysz 39tl, 39bc; ULADZIMIR CYARGEENKA 129br; Valentyn Volkov 22br; Viachaslau Krasnou 48bc; Viktoriia Chursina 53bl; Viktoriia Panchenko 12cl; Visharo 20br; vivoo 36tr; Volodymyr Vorona 110bl; WILDLIFE GmbH 112br; Yllar Hendla 101cl; Yola Watrucka 43tr; Yon Marsh Pipdesigns 105tr; Zbynek Pospisil 36bl.

Dorling Kindersley: 123RF.com: xalanx 60tl; Alan Buckingham 20cl, 24bc, 31br, 34tr, 35tr, 35bl, 35bc, 35br, 38bl, 38bc, 40bl, 41tr, 41bl, 41bc, 54cr, 59tr, 62cr, 63tr, 66br, 68bl, 75tr, 75bl, 76cr, 77tl, 77br, 78cl, 79bc, 102bc, 102br, 104br, 109tl, 110cr, 114br, 116br, 122bl, 131cr, 131bc; Alan Buckingham / Hampton Court Flower Show 2009 59bl; Dreamstime.com: Floriankittemann 41tc; Mark Winwood / RHS Wisley 24tr, 72cr, 86tr; Peter Anderson / RHS Hampton Court Flower Show 2010 70br; Tatton Park 20c.

GAP Photos: 30cl, 31tr, 54bl, 98cl, 137br; Andrea Jones - Design: Adam Frost, Built by Landform Consultants Ltd, Sponsor: Homebase 15br; Claire Higgins 69tr, 83r, 123tl; Clive Nichols 120br; Elke Borkowski - Design Nicola Harding Interior and Garden Design, Sponsor Big Green Egg 139c; Friedrich Strauss 19tr, 128cr; Gary Smith 123bl; Graham Strong 60cr; Jo Whitworth 49c, 117bl, 137cl; John Glover 48br; John Swithinbank 26br; Jonathan Buckley 45bc; Keith Burdett 45tr; Mark Bolton 66c; Mark Bolton - The M G Garden designed by Bunny Guiness at Chelsea Flower Show 2011 17c; Rebecca Bernstein - Design: Gordon Beale 116tr; Robert Mabic 10cr, 37bl; Tim Gainey 71tr; Visions 18cl, 121tl; Visions Premium 13c.

Getty Images: Albert Pi Soler 85tr; Alter_photo 95tl; Andrii Yalanskyi 129bl; art Photo 92bl; asadykov 8bl; benedek 14tr; Birte Gernhardt 104c; Caia Image 10bl; Foodcollection RF 99tl; Funtay 50c; Ivan Marjanovic 121br; Jacky Parker Photography 43bl, 65tl; jirkaejc 97br; joannatkaczuk 10tr; lzf 96cr; Mian Condro 53tr; nitrub 118c; Nordroden 111tr; Peter Cade 29cr; SafakOguz 55tl; SolStock 93tl; Tatsiana Volkava 126br; victorrn 18tr.

Illustrations by Debbie Maizels

All other images © Dorling Kindersley

Produced for DK by
COBALT ID
www.cobaltid.co.uk

Managing Editor Marek Walisiewicz
Editors Diana Loxley, George Arthurton
Managing Art Editor Paul Reid
Art Editor Darren Bland

DK LONDON

Project Editor Amy Slack
Editor Lucy Sienkowska
Senior Designer Glenda Fisher
Managing Editor Ruth O'Rourke
Managing Art Editor Marianne Markham
Production Editor David Almond
Production Controller Stephanie McConnell
Jacket Designer Amy Cox
Jacket Co-ordinator Jasmin Lennie
Art Director Maxine Pedliham
Publisher Katie Cowan

First published in Great Britain in 2023 by
Dorling Kindersley Limited
DK, One Embassy Gardens, 8 Viaduct Gardens,
London, SW11 7BW

Copyright © 2023 Dorling Kindersley Limited
A Penguin Random House Company
23 24 25 26 27 10 9 8 7 6 5 4 3 2 1
001–333475–Jan/2023

A CIP catalogue record for this book is available from the British Library.
ISBN: 978-0-2415-9326-4

Printed and bound in China

For the curious
www.dk.com

MIX
Paper | Supporting responsible forestry
FSC™ C018179